Theories of Probability

An Examination of Logical and Qualitative Foundations

ADVANCED SERIES ON MATHEMATICAL PSYCHOLOGY

Series Editors: H. Colonius (*University of Oldenburg, Germany*)
E. N. Dzhafarov (*Purdue University, USA*)

Vol. 1: The Global Structure of Visual Space
by T. Indow

Vol. 2: Theories of Probability: An Examination of Logical and Qualitative Foundations
by L. Narens

Advanced Series on Mathematical Psychology Vol. 2

Theories of Probability

An Examination of Logical and Qualitative Foundations

Louis Narens
University of California, Irvine, USA

NEW JERSEY • LONDON • SINGAPORE • BEIJING • SHANGHAI • HONG KONG • TAIPEI • CHENNAI

Published by

World Scientific Publishing Co. Pte. Ltd.
5 Toh Tuck Link, Singapore 596224
USA office: 27 Warren Street, Suite 401-402, Hackensack, NJ 07601
UK office: 57 Shelton Street, Covent Garden, London WC2H 9HE

Library of Congress Cataloging-in-Publication Data
Narens, Louis.
Theories of probability : an examination of logical and qualitative foundations / Louis Narens.
p. cm. -- (Advanced series on mathematical psychology ; v. 2)
Includes bibliographical references and index.
ISBN-13: 978-981-270-801-4
ISBN-10: 981-270-801-4
1. Probabilities. I. Title.
QA273.4.N37 2007
519.2--dc22

2007018216

British Library Cataloguing-in-Publication Data
A catalogue record for this book is available from the British Library.

Printed in Singapore.

For Kimberly

Contents

Theories of Probability

an Examination of Logical and Qualitative Foundations

LOUIS NARENS
University of California, Irvine

Chapter 1

Elementary Concepts

1.1 Introduction

Traditional probability theory is founded on Kolmogorov's (1933) axiomatization of a probability function, which assumes probability is a σ-additive measure. This allows for the powerful and highly developed mathematics of measure theory to be immediately available as part of the theory. It is argued in this book, as well as in many places in the literature, that the measure theoretic foundation, while widely applicable, is overspecific for a general concept of probability. This book proposes two different approaches to a more general concept.

The first is qualitative. Kolmogorov's axiomatization assumes numbers (probabilities) have been assigned to events, and his axioms involve both properties of numbers and events. But where did the numbers come from? Some have tried to answer this by having probabilistic assignments be determined by some rule involving random processes. For example, in von Mises (1936) probabilities are limits of relative frequencies arising from random sequences. Obviously approaches based on randomness are limited to situations where assumptions about randomness are appropriate for the generation of the kind of uncertainty under consideration. It is unlikely, for example, that such assumptions apply to the kind of uncertain events encountered in everyday situations. The qualitative approach introduces numbers (probabilities) without making assumptions about randomness. It assumes that *some pairs* of events are comparable in terms of their likelihood of occurrence; that is, some pairs of events are comparable through the relation $\precsim$, where $A \precsim B$ stands for "A is less or equally likely to occur as B." Qualitative axioms are given in terms of events and the relation $\precsim$ that guarantee the existence of a function φ on events such that for the sure event, X, the null event, $\varnothing$, and all A and B in the domain of φ,

(*i*) φ is into the close interval $[0, 1]$ of the reals, $\varphi(X) = 1$, and $\varphi(\varnothing) = 0$,

(*ii*) if $A \cap B = \varnothing$, then $\varphi(A \cup B) = \varphi(A) + \varphi(B)$, and

(*iii*) if $A \precsim B$, then $\varphi(A) \leq \varphi(B)$.

This qualitative approach yields a more general theory than Kolmogorov's, and it applies to important classes of probabilistic situations for which Kolmogorov's axiomatization is overspecific. Additional qualitative axioms can be added so that φ satisfies Kolmogorov's axioms.

I view this book's axiomatic, qualitative approach as being essentially about the same kind of uncertainty covered by the Kolmogorov axiomatization. This kind of uncertainty is one dimensional in nature and is measurable through probability functions or a modest generalization of them. The second approach is about a different kind of uncertainty.

In the decision theory literature, many have suggested that the utility of a gamble involving uncertain events is not its expectation with respect to utility of outcomes, but a more complicated function involving utility of outcomes, subjective probabilities, *and other factors of uncertainty*, for example, knowledge or hypotheses about the processes giving rise to the uncertainty inherent in the events. I find it reasonable to suppose that uncertainty with such "other factors" give rise to a subjective belief function that does not necessarily have properties (*i*) and (*ii*) above of a Kolmogorov probability function. In the models presented in the book, the "other factors" impact belief in two different, but related, ways: (1) by distorting in a systematic manner a Kolmogorov probability function to produce a non-additive belief function (i.e., a belief function $\mathbb{B}$ such that $\mathbb{B}(A \cup B) \neq \mathbb{B}(A) + \mathbb{B}(B)$ for some disjoint events A and B); and (2) by changing the nature of the event space so that it is no longer properly modeled as a boolean algebra of events. Quantum mechanics employs (2) in its modeling of uncertainty. This book's implementation of (2) uses a different kind of event space than those found in quantum mechanics. However, as in quantum mechanics, the belief functions for these event spaces retain abstract properties similar to those of a Kolmogorov probability function. In particular, generalized versions of (*i*) and (*ii*) above are retained.

The book's two approaches can be read separately using the following plan:

Qualitative Foundation: Chapters 1 to 5 and 11.

New Event Space: Chapters 1 and 8 to 10.[1]

[1]One proof in Chapter 9 use concepts of Chapter 4.

Chapters 6 and 7 can be added to either plan. Chapter 7 (which depends on Chapter 6) provides a qualitative foundation for a descriptive theory of human probability judgments known as Support Theory. It employs a boolean event space and axiomatizes a belief function that has a more generalized form than a Kolmogorov probability function. A different foundation for Support Theory is given in Chapter 10. It is based on a non-boolean event space.

The book is not intended to be comprehensive. Much of its material comes from articles by the author. The good part of such a limitation is that it makes for a compact book with unified themes and methods of proof. The bad part is that many excellent results of the literature are left out.

The book is self-contained. The mathematics in it is at the level of upper division mathematics courses taught in the United States. However, many of its concepts are abstract and require mathematical sophistication and abstract thinking beyond that level, but not beyond what is usually achieved by researchers in applied mathematical disciplines like theoretical physics, theoretical computer science, philosophical logic, theoretical economics, etc.

1.2 Preliminary Conventions and Definitions

Convention 1.1 Throughout the book, the following notation, conventions and definitions are observed:

$\mathbb{R}$ denotes the set of reals, $\mathbb{R}^+$ the set of positive reals, $\mathbb{I}$ the integers, $\mathbb{I}^+$ the positive integers, and $*$ the operation of function composition. Usual set-theoretic notation is employed throughout, for example, $\cup$, $\cap$, $-$, and $\in$ are respectively, set-theoretic intersection, union, difference, and membership. $\subseteq$ is the subset relation, and $\subset$ is the proper subset relation, $\varnothing$ is the empty set, and $\wp(A)$ is the power set of A, $\{B|B \subseteq A\}$. $\notin$ stands for "is not a member of" and $\nsubseteq$ for "is not a subset of." For nonempty sets $\mathcal{E}$, $\bigcup\mathcal{E}$ and $\bigcap\mathcal{E}$ have the following definitions:

$$\bigcup\mathcal{E} = \{x|x \in E \text{ for some } E \text{ in } \mathcal{E}\} \text{ and } \bigcap\mathcal{E} = \{x|x \in E \text{ for all } E \text{ in } \mathcal{E}\}.$$

"iff" stands for "if and only if." □

Definition 1.1 Let X be a set. Then X is said to be *denumerable* if and only if there exists a one-to-one function from $\mathbb{I}^+$ onto X. X is said to be *countable* if and only if X is denumerable or X is finite. □

Definition 1.2 Let X be a nonempty set and $\precsim$ be a binary relation on X. Then $\precsim$ is said to be:

Reflexive if and only if for all x in X, $x \precsim x$.

Transitive if and only if for all x, y, and z in X, if $x \precsim y$ and $y \precsim z$ then $x \precsim z$.

Symmetric if and only if for all x and y in X, if $x \precsim y$ then $y \precsim x$.

Connected if and only if for all x and y in X, either $x \precsim y$ or $y \precsim x$.

Antisymmetric if and only if for all x and y in X, if $x \precsim y$ and $y \precsim x$, then $x = y$.

The binary relations $\prec$, $\succsim$, $\succ$, and $\sim$ are defined in terms of $\precsim$ as follows: For all x and y in X,

$x \prec y$ if and only if $x \precsim y$ and not $y \precsim x$.

$x \succsim y$ if and only if $y \precsim x$.

$x \succ y$ if and only if $y \prec x$.

$x \sim y$ if and only if $x \precsim y$ and $y \precsim x$. □

Definition 1.3 Let $\precsim$ be a binary relation on the nonempty set X. Then $\precsim$ is said to be a:

Partial ordering on X if and only if X is a nonempty set and $\precsim$ is a reflexive, transitive, and antisymmetric relation on X.

Weak ordering if and only if $\precsim$ is transitive and connected.

Total ordering if and only if $\precsim$ is a weak ordering and is antisymmetric.

It is immediate that weak and total orderings are reflexive. By convention, partial orderings and total orderings $\precsim$ are often written as $\preceq$ to emphasize the fact that the relation $\sim$ defined in terms of $\precsim$ is the identity relation $=$. □

Definition 1.4 $\equiv$ is said to be an *equivalence relation* on X if and only if X is a nonempty set and $\equiv$ is a reflexive, transitive, and symmetric relation on X. □

It easily follows that if $\precsim$ is a weak odering on X, then $\sim$ is an equivalence relation on X.

The following definition is useful for distinguishing the usual total ordering of the real numbers from the usual total ordering of the rational numbers.

Definition 1.5 Suppose $\preceq$ is a total ordering on X. Then (A, B) is said to be a *Dedekind cut* of $\langle X, \preceq \rangle$ if and only if

(*i*) A and B are nonempty subsets of X,

(*ii*) $A \cup B = X$, and

(*iii*) for each x in A and each y in B, $x \prec y$.

Suppose (A, B) is a Dedekind cut of $\langle X, \preceq \rangle$, where $\preceq$ is a total ordering on X. Then c is said to be a *cut element* of (A, B) if and only if either

(1) c is in A and $x \preceq c \prec y$ for each x in A and each y in B, or

(2) c is in B and $x \prec c \preceq y$ for each x in A and each y in B.

$\langle X, \preceq \rangle$ is said to be *Dedekind complete* if and only if each Dedekind cut of $\langle X, \preceq \rangle$ has a cut element. □

The following theorem is well-known.

Theorem 1.1 *$\langle \mathbb{R}, \leq \rangle$ is Dedekind complete, and for each Dedekind cut (A, B) of $\langle \mathbb{R}, \leq \rangle$, if r and s are cut elements of (A, B), then $r = s$.* □

Definition 1.6 $A_1, \ldots, A_n$ is said to be a *partition of* X if and only if n is an integer ≥ 2, $A_1, \ldots, A_n$ are nonempty and pairwise disjoint and

$$A_1 \cup \cdots \cup A_n = X . \quad \square$$

Let $\mathcal{P} = A_1, \ldots, A_n$ be a partition of X. Note that by Definition 1.6, X is nonempty, $\varnothing$ is not an element of $\mathcal{P}$, and $\mathcal{P}$ has at least two elements.

A frequently employed principle of set theory is the Axiom of Choice. This axiom is often needed in mathematics to show the existence of various set-theoretic objects. In this book, a well-known equivalent of the Axiom of Choice, called "Zorn's Lemma," is sometimes used in proofs.

Definition 1.7 (Axiom of Choice) *For each nonempty set $\mathcal{Y}$ of nonempty sets there exists a function f on $\mathcal{Y}$ such that for each A in $\mathcal{Y}$, $f(A) \in A$.* □

Definition 1.8 Suppose $\mathcal{Y}$ is a nonempty set of sets. Then $A \in \mathcal{Y}$ is said to be *a maximal element of* $\mathcal{Y}$ *with respect to* $\subseteq$ if and only if for each B in $\mathcal{Y}$, if $A \subseteq B$ then $A = B$. □

Definition 1.9 $\mathcal{Y}$ is said to be a *chain* if and only if $\mathcal{Y}$ is a nonempty set of sets and for all A and B in $\mathcal{Y}$, either $A \subseteq B$ or $B \subseteq A$. □

Definition 1.10 (Zorn's Lemma) *Suppose $\mathcal{Y}$ is a nonempty set of sets such that for each subset $\mathcal{Z}$ of $\mathcal{Y}$, if $\mathcal{Z}$ is a chain then $\bigcup \mathcal{Z}$ is in $\mathcal{Y}$. Then $\mathcal{Y}$ has a maximal element with respect to $\subseteq$.* $\square$

Definition 1.11 $\mathcal{F}$ is said to be a *ratio scale* family of functions if and only if $\mathcal{F}$ is a nonempty set of functions from some nonempty set into $\mathbb{R}^+$ such that (i) rf is in $\mathcal{F}$ for each r in $\mathbb{R}^+$ and each f in $\mathcal{F}$, and (ii) for all g and h in $\mathcal{F}$, there exists s in $\mathbb{R}^+$ such that $g = sh$. $\square$

Convention 1.2 In Definition 1.11, "ratio scale" is defined for a family of functions that are into $\mathbb{R}^+$. Occasionally, this concept of "ratio scale" needs to be expanded to include cases where the elements of $\mathcal{F}$ are into $\mathbb{R}^+ \cup \{0\}$ while satisfying the rest of Definition 1.11. The expanded concept is also called a "ratio scale." When the context does not make clear which concept of "ratio scale" is involved, the concept in Definition 1.11 should be used. $\square$

Definition 1.12 Then $\mathcal{F}$ is said to be an *interval scale* family of functions if and only if $\mathcal{F}$ is a nonempty set of functions from some nonempty set into $\mathbb{R}$ such that (i) $rf + s$ is in $\mathcal{F}$ for each r in $\mathbb{R}^+$, each s in $\mathbb{R}$, and each f in $\mathcal{F}$, and (ii) for all g and h in $\mathcal{F}$, there exist q in $\mathbb{R}^+$ and t in $\mathbb{R}$ such that $g = qh + t$. $\square$

Convention 1.3 The notation (a, b) will often stand for the ordered pair of elements a and b, and in general $(a_1, \ldots, a_n)$ will stand for the ordered n-tuple of elements $a_1, \ldots, a_n$. The notation $\langle a_1, \ldots, a_n \rangle$ will also be used to stand for the ordered n-tuple of elements $a_1, \ldots, a_n$. $\langle \cdots \rangle$ is usually used to describe *relational structures with finitely many primitives.* These structures have the form

$$\mathfrak{A} = \langle A, R_1, \ldots, R_m, a_1, \ldots, a_n \rangle ,$$

where A is a nonempty set, $R_1, \ldots, R_m$ are relations on A, and $a_1, \ldots, a_n$ are elements of A. $A, R_1, \ldots, R_m, a_1, \ldots, a_n$ are called the *primitives* of $\mathfrak{A}$. $\square$

Definition 1.13 Let R be an n-ary relation and A be a set. Then the *restriction of R to A,* in symbols, $R \upharpoonright A$, is

$$\{(a_1, \ldots, a_n) \,|\, a_1 \in A, \ \ldots, \ a_n \in A, \ \text{and} \ R(a_1, \ldots, a_n)\} . \ \square$$

Convention 1.4 The convention of mathematics is often employed of having the same symbol denote different relations when a structure and substructure are simultaneously considered, for example, $+$ denoting addition of positive integers in $\langle \mathbb{I}^+, + \rangle$ as well as addition of real numbers in $\langle \mathbb{R}, + \rangle$. $\square$

Chapter 2

Kolmogorov Probability Theory

Since the 1930's, the probability calculus of Kolmogorov (1933, 1950) has become the standard theory of probability for mathematics and science. Many philosophers of science and statisticians consider it to be the foundation for a general, rational theory of belief involving uncertainty. The author and others have been critical of this view and consider it to be a theory of probability that is at best only rationally justifiable in certain narrow kinds of probabilistic situations, for example, continuous situations in physics. That is, we believe the Kolmogorov theory is overspecific for a general, rational theory of belief.

The Kolmogorov theory assumes the following definition and six axioms:

Definition 2.1 $\mathcal{A}$ is said to be a *boolean algebra of subsets* of X if and only if the following five conditions hold:

(1) X is a nonempty set and $\mathcal{A}$ is a set of subsets of X;

(2) X is in $\mathcal{A}$, and the empty set, $\varnothing$, is in $\mathcal{A}$;

(3) for all A and B, if A is in $\mathcal{A}$ and B is in $\mathcal{A}$, then $A \cap B$ is in $\mathcal{A}$;

(4) for all A and B, if A is in $\mathcal{A}$ and B is in $\mathcal{A}$, then $A \cup B$ is in $\mathcal{A}$; and

(5) for all A in $\mathcal{A}$, $X - A$ is in $\mathcal{A}$.

$\mathcal{A}$ is said to be a *boolean σ-algebra of subsets* if and only if $\mathcal{A}$ is a boolean algebra of subsets such that if $A_i \in \mathcal{A}$ for each $i \in \mathbb{I}^+$, then $\bigcup_{i=1}^{\infty} A_i$ is in $\mathcal{A}$.

Suppose $\mathcal{A}$ is a boolean algebra of sets. Then $\mathcal{B}$ is said to be a *subalgebra* of $\mathcal{A}$ if and only if $\mathcal{B} \subseteq \mathcal{A}$ and $\mathcal{B}$ is a boolean algebra of sets. □

The following six axioms summarize Kolmogorov's axiomatic treatment of probability.

Axiom 2.1 *Uncertainty is captured by a unique function* $\mathbb{P}$. □

Axiom 2.2 *The domain of* $\mathbb{P}$ *is a boolean* σ*-algebra of subsets (Definition 2.1).* □

Axiom 2.3 *The codomain of* $\mathbb{P}$ *is a subset of the closed interval of real numbers* $[0, 1]$. □

Axiom 2.4 $\mathbb{P}(\varnothing) = 0$. □

Axiom 2.5 $\mathbb{P}(X) = 1$. □

Axiom 2.6 (σ-additivity) *If* A_i, $i \in \mathbb{I}^+$, *is a sequence of pairwise disjoint sets, then*

$$\mathbb{P}(\bigcup_{i=1}^{\infty} A_i) = \sum_{i=1}^{\infty} \mathbb{P}(A_i) \, . \quad \square$$

Definition 2.2 A function $\mathbb{P}$ satisfying Axioms 2.2 to 2.6 is called a *σ-additive probability function (on $\mathcal{A}$).* □

Except for Axiom 2.4, $\mathbb{P}(\varnothing) = 0$, it will be argued at various places in this book that each of the other five Kolmogorov axioms are overspecific. Axiom 2.6, σ-additivity, is generally singled out in the literature as being overspecific, and it is often suggested that to achieve a more general theory of probability, Axioms 2.2 and 2.6 should be replaced by Axioms 2.7 and 2.8 below.

Axiom 2.7 *The domain of* $\mathbb{P}$ *is a boolean algebra of subsets (Definition 2.1).* □

Axiom 2.8 (finite additivity) *For all* A *and* B *in* $\mathcal{A}$, *if* $A \cap B = \varnothing$, *then*

$$\mathbb{P}(A \cup B) = \mathbb{P}(A) + \mathbb{P}(B) \, . \quad \square$$

However, it is argued in this book that Axioms 2.7 and 2.8 are still overspecific.

Definition 2.3 A function $\mathbb{P}$ satisfying Axioms 2.3, 2.4, 2.5, 2.7, and 2.8 is called a *finitely additive probability function (on $\mathcal{A}$).* □

This book emphasizes the more general situation of finitely additive probability functions instead of σ-additive probability functions. By convention, the term, "the Kolmogorov theory," applies to both types of functions.

Convention 2.1 By convention, when the term "probability function" is used without the prefixes "finitely additive" or "σ-additive", it refers to a finitely additive probability function. When σ-additivity is needed, the prefix "σ-additive" is added. □

In the Kolmogorov theory the important probabilistic concepts of conditional probability and independence are defined in terms of $\mathbb{P}$:

Definition 2.4 For all A and B in $\mathcal{A}$ such that $\mathbb{P}(B) \neq 0$, the *conditional probability of A given B*, in symbols, $\mathbb{P}(A|B)$, is defined as

$$\mathbb{P}(A|B) = \frac{\mathbb{P}(A \cap B)}{\mathbb{P}(B)}. \quad \square$$

Definition 2.5 For all A and B in $\mathcal{A}$, A and B are said to be *independent,* in symbols, $A \perp B$, if and only if

$$\mathbb{P}(A \cap B) = \mathbb{P}(A)\mathbb{P}(B). \quad \square$$

Probability theory is enormously applicable. Part of the reason is due to its rich mathematical calculus for manipulating probabilistic quantities. The richness comes from the following correspondence: addition corresponds to forming disjoint unions of events, multiplication to the intersection of independent events, and division to conditioning one event on another.

Many alternatives to the Kolmogorov theory have been proposed in the literature. Some are generalizations; others are about a different kind of "probability." With few exceptions, the numerical assignments of the alternatives form very weak calculi of quantities. In contrast, this book presents alternatives that have calculi that rival the finitely additive version of the Kolmogorov theory in terms of mathematical richness, and in some cases exceed it.

The main competitor in the literature to Kolmogorov's (1933) theory has been the relative frequency approach of Richard von Mises (1936), where probabilities are defined as limits of sequences of relative frequencies of random events. Although many textbooks give limiting relative frequencies as the definition of probability, its rigorous development is almost never attempted in those books, which in addition often fail to mention that probability functions that are produced in this manner are finitely additive, and not σ-additive.[1]

[1] Descriptions and critical evaluations of prominent approaches to probability theory can be found in Terrence Fine's excellent book, *Theories of Probability* (Fine, 1973).

This book pursues very different foundational approaches to probability theory than those of Kolmogorov and von Mises. One is based on a strategy developed by various behavioral and economic scientists and philosophers. It assumes an ordering, $\precsim$, on a set of events $\mathcal{E}$. "$A \precsim B$" is usually read as "the event A is less or equally likely to occur than the event B." The measurement problem for this kind of situation is showing $\langle \mathcal{E}, \precsim \rangle$ has a *probability representation,* that is, showing the existence of a finitely additive probability function $\mathbb{P}$ on $\mathcal{E}$ such that

$$\text{if } A \precsim B \text{ then } \mathbb{P}(A) \leq \mathbb{P}(B)\,. \tag{2.1}$$

When Equation 2.1 holds, it is often said that "$\mathbb{P}$ represents $\precsim$." At this level of analysis, the qualitative theory is more general than the finitely additive version of the Kolmogorov theory, because it does not necessarily produce a unique probability function for representing $\precsim$. Nevertheless, as is shown in Chapter 4, it is still a mathematical rich probability theory. Some researchers, including the author, consider the lack of uniqueness to be an important generalization of the Kolmogorov theory.

Many researchers of probability have developed theories to represent strengths of personal belief as Kolmogorov probabilities. They often provide arguments that claim the rational assignment of numbers to beliefs must obey the Kolmogorov axioms. I and others consider the Kolmogorov theory to be overspecific for many belief situations. We believe it needs to be extended. (An extension that encompasses additional rational phenomena is proposed in Chapter 9; extensions that encompass human judgments of probability are presented in Chapters 7 and 10.)

Chapter 3

Infinitesimals

3.1 Introduction

Probability theory is one of several interpretations of measure theory. In it the set X is interpreted as a sample space consisting of the set of possible states world, the boolean algebra $\mathcal{E}$ of subsets of X as a set of events, and the measure $\mathbb{P}$ as a function that assigns to each event A in $\mathcal{E}$ the probability that the actual state of the world is in A. Although each x in X is considered to have some chance of occurring, many probabilistic situations are modeled in a manner such that $\mathbb{P}(\{x\}) = 0$. In such cases, the possibility of the occurrence of an element of X is not distinguishable in terms of probability from the impossibility of the occurrence of the impossible event $\varnothing$. The inability to make this distinction rules out many natural concepts for dealing with events of probability 0. Conditioning on events of probability 0 is an example: Consider the case where $\mathbb{P}$ arises from a uniform distribution on an infinite set X and x and y are elements of X. Then one would want $\mathbb{P}(\{x\} \mid \{x, y\}) = .5$. The obvious and natural way of extending probability theory to provide for this, and more generally for a more structured approach to events of probability 0, is to have the co-domain of $\mathbb{P}$ include infinitesimal quantities. As is shown in Chapter 4, such an inclusion not only provides a closer match to the intuitive concept of "chance of occurring," but also provides methods that often make the mathematics of probabilistic situations much easier to deal with—even when infinitesimals are not mentioned as part of the final theorems. It also provides a more encompassing theory: Results of Chapters 4 and 11 show that the inclusion of infinitesimals provide for sharper qualitative axiomatizations, a better fit with techniques of mathematical logic, and a better foundation for philosophical issues concerning probabilities.

Since the beginnings of calculus in the 17th century to nearly the end of the 19th century, infinitesimals were widely used in mathematics. They were controversial because of issues involving their ontology and proper use. Ultimately, as the real numbers and concepts of infinity received a rigorous development in mathematics in the later part of the 19th century, they became an embarrassment to rigorously minded mathematicians, because no one knew how to put them, or the practices involving them, on rigorous bases. Other methods were developed to eliminate their need in proofs. This changed in the 1960's when Abraham Robinson developed a rigorous version of the infinitesimal calculus. He constructed infinitesimal quantities through the use of mathematical logic and used them much like the mathematicians of the 17th and 18th centuries. He produced a series of articles in which he applied his version of infinitesimal analysis to a wide variety of mathematical topics. Robinson showed that his version of infinitesimal analysis contained a far richer and more subtle theory of infinitesimal and infinite quantities than the earlier theories and methods. His work in this area was collected together and synthesized in his 1966 book, *Nonstandard Analysis*.

This chapter takes an algebraic approach to infinitesimals. Its method of producing infinitesimals is similar in some respects to classical methods of comparing rates of divergence for infinite series (e.g., G. H. Hardy's *Orders of Infinity*) and earlier algebraic attempts at an infinitesimal calculus (e.g., Schmieden and Langwitz, 1958). Through a model-theoretic result of Łoś (1955), it can be developed to have the full power of Robinson's nonstandard analysis.

The underlying idea used in the chapter for constructing infinitesimals is simple. Consider sequences of real numbers as quantities. Identify constant sequences with real numbers. Identify other sequences with a new kind of quantity, which sometimes will be infinite (with respect to the reals) and other times be infinitesimally close to a real. When such a quantity is infinitesimally close to 0, it is considered to be an "infinitesimal."

In particular, for each real number r, identify the constant sequence

$$f_r = r, r, \ldots, r, \ldots$$

with the real number r. Consider the sequence

$$f = 1, \frac{1}{2}, \frac{1}{3}, \ldots, \frac{1}{n}, \ldots,$$

where n is the nth positive integer. Then for each positive real number r,

$$f_r(i) > f(i) > f_0(i),$$

for all but finitely many i in $\mathbb{I}^+$. Thus if we consider f as representing a number, then it is reasonable to consider f to be smaller than each positive real but larger than 0, that is, f to be a "positive infinitesimal." This is essentially the approach to infinitesimals that is taken in this book.

Definition 3.1 Throughout the remainder of this chapter, let $\mathcal{R}$ be the set of real valued sequences, that is, let

$$\mathcal{R} = \{f|f : I^+ \to \mathbb{R}\}.$$

For each r in $\mathbb{R}$, let f_r be the constant sequence of value r.

By definition, f and g in $\mathcal{R}$ are said to be *equivalent,* in symbols, $f \sim g$, if and only if for all but finitely many i in $\mathbb{I}^+$, $f(i) = g(i)$. It is easy to verify that $\sim$ is an equivalence relation of $\mathcal{R}$ and thus partitions $\mathcal{R}$ into a set of $\sim$-equivalence classes. By definition, for each f in $\mathcal{A}$, let $f^\sim$ be the $\sim$-equivalence class of $\mathcal{R}$ to which f belongs, and ${}^\star\mathbb{R}$ be the set of the equivalences classes of $\mathcal{R}$, that is,

$${}^\star\mathbb{R} = \{f^\sim|f \in \mathcal{R}\}.$$

Because f and g agree on almost all i in $\mathbb{I}^+$, we often write,

$$f = g \text{ a.e.},$$

in words, "f and g are identical almost everywhere," for $f \sim g$. □

Convention 3.1 By convention, each real number r is often identified with the equivalence class $f_r^\sim$ (and more informally with the function f_r). In fact "$f_r^\sim$" will often be written as "r". Thus by these conventions, $\mathbb{R}$ is often considered a subset of ${}^\star\mathbb{R}$. □

Convention 3.1 is justified by a later theorem showing that the algebraic system of real numbers is isomorphic to an appropriately defined algebraic subsystem of ${}^\star\mathbb{R}$.

Definition 3.2 Addition and multiplication are extended to ${}^\star\mathbb{R}$ as follows: For all α, β, and γ in ${}^\star\mathbb{R}$, $\alpha + \beta = \gamma$ if and only if for some $f \in \alpha$, $g \in \beta$, and $h \in \gamma$,

$$f + g = h \text{ a.e.},$$

and for all α', β', and γ' in ${}^\star\mathbb{R}$, $\alpha' \cdot \beta' = \gamma'$ if and only if for some $f' \in \alpha'$, $g' \in \beta'$, and $h' \in \gamma'$,

$$f' \cdot g' = h' \text{ a.e.} \quad \square$$

It is easy to verify that the just above definitions are independent of the particular elements of α, β, γ, α', β', γ' chosen; that is, if a, b, c, a', b', and c' are such that $a \in \alpha$, $b \in \beta$, $c \in \gamma$, $a' \in \alpha'$, $b' \in \beta'$, and $c \in \gamma'$, then

$$a + b = f + g = h \text{ a.e.}$$

and

$$a' \cdot b' = f' \cdot g' = h' \text{ a.e.}$$

Note that if r, s, and t are reals, then

$$f_r^\sim + f_s^\sim = f_t^\sim \text{ iff } r + s = t\,,$$

and

$$f_r^\sim \cdot f_s^\sim = f_t^\sim \text{ iff } r \cdot s = t\,.$$

Thus if for each real number u, $f_u^\sim$ is identified with u, then addition and multiplication on ${}^\star\mathbb{R}$ becomes an extension of addition and multiplication on $\mathbb{R}$.

A precise and unambiguous treatment of addition and multiplication on ${}^\star\mathbb{R}$ would require the introduction of new symbols, say ${}^\star+$ for addition and ${}^\star\cdot$ for multiplication on ${}^\star\mathbb{R}$, because $+$ and $\cdot$ are already used for addition and multiplication on $\mathbb{R}$. However, the introduction of such symbols often make simple equations difficult to understand. Thus I prefer, for the sake of readability, to occasionally allow such slight ambiguities to slip into the notation.

Definition 3.3 Elements of ${}^\star\mathbb{R}$ are ordered as follows: For all α and β in ${}^\star\mathbb{R}$, $\alpha \leq \beta$ if and only if for some f in α and g in β,

$$f \leq g \text{ a.e.} \quad \square$$

Again, it is easy to verify that the definition of $\leq$ above is independent of the particular elements of α and β chosen.

It is easy to verify that the structure $\langle {}^\star\mathbb{R}, \leq, +, \cdot \rangle$ share many algebraic properties of the real number system, $\langle \mathbb{R}, \leq, +, \cdot \rangle$, for example, the associative laws for addition and multiplication ($(x + y) + z = x + (y + z)$ and $(x \cdot y) \cdot z = x \cdot (y \cdot z)$), the distribution of multiplication over addition ($x \cdot (y + z) = x \cdot y + x \cdot z$), the existence of multiplicative and additive identities ($x + 0 = x$ and $1 \cdot x = x$), and the existence of additive inverses (there exists y such that $x + y = 0$). It is also easy to verify that $f^\sim$, where

$$f(i) = \frac{1}{i}\,,$$

is a positive infinitesimal, that is, for each r in $\mathbb{R}^+$,

$$0 = f_0^\sim < f < f_r^\sim = r\,.$$

However, $\langle {}^\star\mathbb{R}, \leq, +, \cdot\rangle$ has certain algebraic properties that are undesirable for manipulating quantities like probabilities. Consider the two sequences,

$$g = 1, 0, 1, 0, 1, 0, \ldots \text{ and } h = 0, 1, 0, 1, 0, 1, \ldots\,.$$

Then,

$$g^\sim \neq 0,\ h^\sim \neq 0, \text{ and } g^\sim \cdot h^\sim = 0,$$

(i.e., $\langle {}^\star\mathbb{R}, \leq, +, \cdot\rangle$ has "divisors of 0"). Also,

$$\text{neither } g^\sim \leq h^\sim \text{ nor } h^\sim \leq g^\sim,$$

(i.e., the partial ordering $\leq$ on ${}^\star\mathbb{R}$ is not connected). In the following section, a method is provided for eliminating these deficiencies and thereby produce a variant of $\langle {}^\star\mathbb{R}, \leq, +, \cdot\rangle$ that is much more similar algebraically to $\langle \mathbb{R}, \leq, +, \cdot\rangle$.

3.2 Ultrafilters

The basic idea of the construction of ${}^\star\mathbb{R}$ in the previous section was to consider two real-valued sequences "equal" if they agreed almost everywhere, and to define addition, multiplication, and ordering of sequences using the "almost everywhere" relation. In that construction, "almost everywhere" meant "for all but finitely many indices." A strategy for obtaining a better algebraic structure with infinitesimals is to consider other "almost everywhere" relations. To do this, it is convenient to axiomatize the relevant conditions an "almost everywhere" relation should have. This is done in the following definition, where the X is "everywhere" and $\mathcal{F}$ is the set of "almost everywhere" subsets of X.

Definition 3.4 $\mathcal{F}$ is said to be a *filter* on X if and only if X is a nonempty set and the following three conditions hold for all subsets A and B of X:

(1) $X \in \mathcal{F}$ and $\varnothing \notin \mathcal{F}$;

(2) if $A \subseteq B$ and $A \in \mathcal{F}$ then $B \in \mathcal{F}$; and

(3) if $A \in \mathcal{F}$ and $B \in \mathcal{F}$ then $A \cap B \in \mathcal{F}$. $\square$

Many kinds of different kinds of filters exist on infinite sets. The particular filter on $\mathbb{I}^+$ used in the previous section is an example of a co-finite filter on $\mathbb{I}^+$:

Definition 3.5 Let X be an infinite set and $\mathcal{F}$ the set of all *co-finite subsets* of X, that is, let $\mathcal{F}$ be the set of all subsets A of X such that $X - A$ is finite. Then it is easy to verify that $\mathcal{F}$ is a filter on X. $\mathcal{F}$ is called the *filter of co-finite subsets* on X. $\square$

Definition 3.6 Let X be a set and A be a nonempty subset of X. Then

$$\mathcal{F} = \{B \,|\, A \subseteq B \subseteq X\}$$

is a filter on X. $\mathcal{F}$ is called the *principal filter on X generated by A*. $\square$

Definition 3.7 $\mathcal{F}$ is said to be a *non-principal filter* on a nonempty set X if and only if $\mathcal{F}$ is a filter on X and $\bigcap \mathcal{F} = \varnothing$. $\square$

Let X be a nonempty set. It easily follows that a principal filter on X is not non-principal. Let $\mathcal{F}$ be the set of co-finite subsets of X. Because for each $x \in X$, $X - \{x\}$ is a co-finite subset of X, it follows that $\bigcap \mathcal{F} = \varnothing$. Thus the co-finite filter on X is non-principal. As the reader can easily verify, if there is a non-principal filter on X, then X is infinite.

The following property is useful for constructing filters.

Definition 3.8 $\mathcal{F}$ is said to have the *finite intersection property* if and only if $\mathcal{F}$ is a nonempty set of subsets and each nonempty finite subset $\{A_1, \ldots, A_n\}$ of $\mathcal{F}$,

$$A_1 \cap \ldots \cap A_n \neq \varnothing\,. \quad \square$$

Theorem 3.1 *Suppose that X is a nonempty set and $\mathcal{F}$ is a set of subsets of X that has the finite intersection property. Then there exists a filter $\mathcal{G}$ on X such that $\mathcal{F} \subseteq \mathcal{G}$.*

Proof. Let

$$\begin{aligned}\mathcal{G} \;=\; & \{A | A \subseteq X \text{ and for some } n \in \mathbb{I}^+ \text{ there exist} \\ & A_1, \ldots, A_n \text{ in } \mathcal{F} \text{ such that } A_1 \cap \ldots \cap A_n \subseteq A\}\,.\end{aligned}$$

Then it is easy to verify that $\mathcal{G}$ is a filter. $\square$

The algebra $\langle {}^\star\mathbb{R}, \leq, +, \cdot\rangle$ considered in Section 3.1 lacked certain desirable algebraic characteristics of the real number system $\langle \mathbb{R}, \leq, +, \cdot\rangle$, for example, the connectedness of the ordering $\leq$ on ${}^\star\mathbb{R}$ and having no divisors of 0. As shown a little later in this section, such lacks vanish when the "almost everywhere" relation is defined in terms of a special kind of filter on $\mathbb{I}^+$ called an *ultrafilter*.

Definition 3.9 $\mathcal{F}$ is said to be an *ultrafilter* on X if and only if $\mathcal{F}$ is a filter on X and for all filters $\mathcal{G}$ on X, if $\mathcal{F} \subseteq \mathcal{G}$ then $\mathcal{F} = \mathcal{G}$. □

Theorem 3.2 *For each filter $\mathcal{F}$ on X there exists an ultrafilter $\mathcal{U}$ on X such that $\mathcal{F} \subseteq \mathcal{U}$.*

Proof. Let

$$\Gamma = \{\mathcal{G} | \mathcal{G} \text{ is a filter on } X \text{ and } \mathcal{F} \subseteq \mathcal{G}\}.$$

Then $\Gamma \neq \varnothing$, because $\mathcal{F} \in \Gamma$. If Δ is a chain in Γ (Definition 1.9) then $\bigcup \Delta$ is in Γ. Thus by Zorn's Lemma (Definition 1.10), Γ has a maximal element $\mathcal{U}$ (Definition 1.8), and this maximal element is an ultrafilter. □

Theorem 3.3 *Let $\mathcal{U}$ be a filter on X. Then the following two statements are equivalent:*

1. *$\mathcal{U}$ is an ultrafilter on X.*
2. *For each subset A of X, either $A \in \mathcal{U}$ or $(X - A) \in \mathcal{U}$.*

Proof. Assume 1. Suppose 2 does not hold. A contradiction will be shown. Let A be the subset of X such that neither $A \in \mathcal{U}$ nor $(X - A) \in \mathcal{U}$. Let

$$\mathcal{F} = \{C \,|\, \text{there exists } B \text{ in } \mathcal{U} \text{ such that } A \cap B \subseteq C \subseteq X\}.$$

Then $\mathcal{U} \subseteq \mathcal{F}$. Because $A \in \mathcal{F}$, $\mathcal{U} \subset \mathcal{F}$. Because $\mathcal{U}$ is an ultrafilter, to show a contradiction it needs only be shown that $\mathcal{F}$ is a filter. From the definition of $\mathcal{F}$, it immediately follows that (i) $X \in \mathcal{F}$, and (ii) for each K in $\mathcal{F}$ and each subset L of X, if $K \subseteq L$ then $L \in \mathcal{F}$. Suppose D and E are arbitrary elements of $\mathcal{F}$. Let F and G be elements of $\mathcal{U}$ such that $A \cap F \subseteq D$ and $A \cap G \subseteq E$. Then

$$A \cap (F \cap G) = (A \cap F) \cap (A \cap G) \subseteq D \cap E,$$

and thus, because $F \cap G$ is in $\mathcal{U}$, $D \cap E$ is in $\mathcal{F}$. Finally, $\varnothing$ is not in $\mathcal{F}$, because if it were in $\mathcal{F}$, then for some H in $\mathcal{U}$, $H \cap A = \varnothing$, and thus $H \subseteq X - A$, that is, $X - A$ would be in $\mathcal{U}$, contrary to our assumptions.

Assume 2. Suppose $\mathcal{U}$ is not an ultrafilter. A contradiction will be shown. Let $\mathcal{G}$ be a filter on X such that $\mathcal{U} \subset \mathcal{G}$. Let B be in $\mathcal{G} - \mathcal{U}$. Because $B \notin \mathcal{U}$, it follows by assumption that $X - B$ is in $\mathcal{U}$. Because $\mathcal{U} \subseteq \mathcal{G}$, $X - B$ is in $\mathcal{G}$. Thus both B and $X - B$ are in $\mathcal{G}$, and because $\mathcal{G}$ is a filter, $B \cap (X - B) = \varnothing$ is in $\mathcal{G}$, contradicting that $\mathcal{G}$ is a filter. □

Definition 3.10 Let X be a nonempty set and a be an element of X. Then it is easy to show that

$$\mathcal{U} = \{A | A \subseteq X \text{ and } a \in A\}$$

is an ultrafilter on X. $\mathcal{U}$ is called the *principal ultrafilter on X generated by a*. □

It is easy to show that if $\mathcal{F}$ is an ultrafilter on X and a principal filter on X (Definition 3.6), then for some b in X, $\mathcal{F}$ is the principal ultrafilter on X generated by b.

3.3 Ultrapowers of the Reals

In this section, the construction of $^{\star}\mathbb{R}$ in Section 3.1 is modified in two ways. First, sequences are defined on an arbitrary set of indexes, X, rather than just on $\mathbb{I}^+$. And second, the "almost everywhere" relation $\sim$ is defined with respect to an ultrafilter on X, rather than just the filter of co-finite subsets of $\mathbb{I}^+$. These two modifications produce a much richer theory of infinitesimals.

Definition 3.11 Let $\mathcal{U}$ be an ultrafilter on X and

$$\mathcal{R} = \{f | f : X \to \mathbb{R}\}.$$

Define the binary relation $\sim$ on $\mathbb{R}$ as follows:

$$f \sim g \text{ iff } \{i | f(i) = g(i)\} \in \mathcal{U}.$$

Then it is easy to verify that $\sim$ is an equivalence relation on $\mathcal{R}$. Let $^{\star}\mathbb{R}$ be the set of $\sim$-equivalence classes of $\mathcal{R}$.

For each a in $\mathbb{R}$, define $^{\star}a$ as follows: Let h_a be the function on X such that for all i in X, $h_a(i) = a$. Then $^{\star}a = h_a^{\sim}$. $^{\star}a$ is called the *$\mathcal{U}$-extension* of a.

For each n-ary relation $S(x_1, \ldots, x_n)$ on $\mathbb{R}$, define the n-ary relation $^{\star}S$ on $^{\star}\mathbb{R}$ as follows: For each $\alpha_1, \ldots, \alpha_n$ in $^{\star}\mathbb{R}$,

$$^{\star}S(\alpha_1, \ldots, \alpha_n) \text{ iff for some } f_1 \in \alpha_1, \ldots, f_n \in \alpha_n, \\ \{i | S(f_1(i), \ldots, f_n(i))\} \in \mathcal{U}.$$

$^{\star}S$ is called the *$\mathcal{U}$-extension* of S. Note that $^{\star}\mathbb{R}$ is the $\mathcal{U}$-extension of $\mathbb{R}$. □

The concept of "$\mathcal{U}$-extension" in Definition 3.11 is well-defined in the following sense: Suppose ${}^\star S$ is the $\mathcal{U}$-extension of the n-ary relation S on $\mathbb{R}$ and $\alpha_1, \ldots, \alpha_n$ are such that ${}^\star S(\alpha_1, \ldots, \alpha_n)$. By Definition 3.11, let $f_1, \ldots, f_n$ be such that

$$f_1 \in \alpha_1, \ldots, f_n \in \alpha_n \text{ and } \{i | S(f_1(i), \ldots, f_n(i))\} \in \mathcal{U}.$$

Let $g_1, \ldots, g_n$ be arbitrary elements of, respectively, $\alpha_1, \ldots, \alpha_n$. For $j = 1, \ldots, n$, let

$$A_j = \{i | f_j(i) = g_j(i)\}.$$

Then for $j = 1, \ldots, n$, $A_j \in \mathcal{U}$. Because $\mathcal{U}$ is a filter,

$$A_1 \cap \ldots \cap A_n \in \mathcal{U}.$$

Thus,

$$\{i | S(g_1(i), \ldots, g_n(i))\} \supseteq A_1 \cap \ldots \cap A_n \in \mathcal{U}.$$

${}^\star S$ may be looked at as an extension of the n-ary relation S on $\mathbb{R}$ as follows: For each $r \in \mathbb{R}$, let c_r be the function from X into $\mathbb{R}$ such that for each i in X, $c_r(i) = r$, and let α_r be the element of ${}^\star\mathbb{R}$ such that $c_r \in \alpha_r$. Then

$$\begin{aligned} {}^\star S(\alpha_{r_1}, \ldots, \alpha_{r_n}) \ &\text{iff} \ \{i | S(c_{r_1}(i), \ldots, S(c_{r_n}(i))\} \in \mathcal{U} \\ &\text{iff} \ S(r_1, \ldots, r_n). \end{aligned}$$

Therefore, the function $\varphi : \mathbb{R} \to^\star \mathbb{R}$ such that for each r in $\mathbb{R}$,

$$\varphi(r) = \alpha_r,$$

is an isomorphic imbedding of S into ${}^\star S$.

Convention 3.2 Given the above notation, it has been shown that ${}^\star S$ may be looked at as an extension of S for each n-ary relation S on $\mathbb{R}$. In particular, $\mathbb{R}$ is often viewed as a subset of ${}^\star\mathbb{R}$, resulting in the harmless confusing of the real number r with the equivalence class α_r. □

Operations are special kinds of relations. For example, the binary operation $+$ on $\mathbb{R}$ may be viewed as the ternary relation S on $\mathbb{R}$ such that for all x, y, and z in $\mathbb{R}$,

$$S(x, y, z) \text{ iff } x + y = z.$$

In general, an n-ary operation F on $\mathbb{R}$ can be thought of as an $(n+1)$-ary relation T on $\mathbb{R}$ such that for all $x_1, \ldots, x_n$, and y,

$$T(x_1, \ldots, x_n, y) \text{ iff } F(x_1, \ldots, x_n) = y.$$

Let V be an $(n+1)$-ary relation on $\mathbb{R}$ that is the n-ary function G on $\mathbb{R}$. That is, V is an $(n+1)$-ary relation on $\mathbb{R}$ such that for all $\alpha_1, \ldots, \alpha_n$ in ${}^\star\mathbb{R}$ there exists exactly one β such that $V(\alpha_1, \ldots, \alpha_n, \beta)$, and we write

$$\beta = G(\alpha_1, \ldots, \alpha_n)$$

to describe this. Let $\mathcal{U}$ be an ultrafilter on the set X and let ${}^\star V$ be the $\mathcal{U}$-extension of V. It will be shown that ${}^\star V$ is the n-ary operation ${}^\star G$ on ${}^\star\mathbb{R}$ that is the $\mathcal{U}$-extension of G. Let $\alpha_1, \ldots, \alpha_n$ be arbitrary elements of $\mathbb{R}$ and for $k = 1, \ldots, n$, let $f_k \in \alpha_k$. Because G is an n-ary operation on $\mathbb{R}$, for each i in X let

$$h(i) = G(f_1(i), \ldots, f_n(i)).$$

Then

$$X = \{i | V(f_1(i), \ldots, f_n(i), h(i))\}.$$

Let β be the element of ${}^\star\mathbb{R}$ such that $h \in \beta$. Then ${}^\star V(\alpha_1, \ldots, \alpha_n, \beta)$ holds. Suppose ${}^\star V(\alpha_1, \ldots, \alpha_n, \gamma)$. Let g be an arbitrary element of γ and

$$A = \{i | V(f_1(i), \ldots, f_n(i), g(i))\}.$$

Then $A \in \mathcal{U}$. Thus by the choice of V,

$$\{i | h(i) = g(i)\} = A \in \mathcal{U},$$

that is, $\beta = \gamma$.

Definition 3.12 Let $\mathcal{U}$ be an ultrafilter on X, $S_1, \ldots, S_n$ be relations on $\mathbb{R}$, $a_1, \ldots, a_m$ be elements of $\mathbb{R}$, ${}^\star S_1, \ldots, {}^\star S_n$ be $\mathcal{U}$-extensions of $S_1, \ldots, S_n$, and ${}^\star a_1, \ldots, {}^\star a_n$ be $\mathcal{U}$-extensions of $a_1, \ldots, a_n$. Then

$$\langle {}^\star\mathbb{R}, {}^\star S_1, \ldots, {}^\star S_n, {}^\star a_1, \ldots, {}^\star a_m \rangle$$

is called the *$\mathcal{U}$-ultrapower* of $\langle \mathbb{R}, S_1, \ldots, S_n, a_1, \ldots, a_m \rangle$. $\square$

Convention 3.3 To simplify notation, the convention of writing simple expressions involving *-extensions of the elements of $\mathbb{R}$, the operations $+$, $\cdot$, and $-$, and the relations $\leq$ and $<$ "without the *-symbol" is generally employed. Thus, for example,

$${}^\star 1 \;{}^\star\!+\; \alpha \;{}^\star\!+\; \beta \;{}^\star\!\leq\; \gamma \;{}^\star\!\cdot\; (\beta \;{}^\star\!+\; \alpha \;{}^\star\!+\; \delta)$$

is generally written as

$$1 + \alpha + \beta \leq \gamma \cdot (\beta + \alpha + \delta),$$

or, with the usual convention of expressing multiplication as juxtaposition, as

$$1 + \alpha + \beta \leq \gamma(\beta + \alpha + \delta). \quad \square$$

In Section 3.1, ${}^\star\mathbb{R}$ was constructed using the filter of co-finite subsets of $\mathbb{I}^+$. This produced an algebraic structure $\langle {}^\star\mathbb{R}, \leq, +, \cdot \rangle$ that lacked the ability to manipulate quantities with ease of the real number system, $\langle \mathbb{R}, \leq, +, \cdot \rangle$. As Theorem 3.5 below shows, this deficiency is eliminated if the construction of ${}^\star\mathbb{R}$ uses an ultrafilter containing the filter of co-finite subsets of $\mathbb{I}^+$.

Lemma 3.1 *Suppose $\mathcal{U}$ is an ultrafilter on X and $\langle {}^\star\mathbb{R}, \leq, +, \cdot \rangle$ is the $\mathcal{U}$-ultrapower of $\langle \mathbb{R}, \leq, +, \cdot \rangle$. Then the following eight statements are true about $\langle {}^\star\mathbb{R}, \leq, +, \cdot \rangle$ for all x, y, z, and w in ${}^\star\mathbb{R}$:*

1. (additive commutativity) $x + y = y + x$.
2. (multiplicative commutativity) $x \cdot y = y \cdot x$.
3. (additive associativity) $(x + y) + z = x + (y + z)$.
4. (multiplicative associativity) $(x \cdot y) \cdot z = x \cdot (y \cdot z)$.
5. (distributivity) $x \cdot (y + z) = x \cdot y + x \cdot z$.
6. (additive identity element) *There exists a in ${}^\star\mathbb{R}$ (called the* additive identity element*) such that for all v in ${}^\star\mathbb{R}$, $a + v = v$.*
7. (additive inverse) *There exists v in $\mathbb{R}$ such that $x + v = a$, where a is the additive identity element.*
8. (multiplicative identity element) *There exists b in ${}^\star\mathbb{R}$ (called the* multiplicative identity element*) such that b is not an additive identity element and for all v in ${}^\star\mathbb{R}$, $b \cdot v = v$.*

Proof. (In this proof only filter properties of $\mathcal{U}$ are needed.) Let α be an arbitrary element of ${}^\star\mathbb{R}$ and let α_0 be the element of ${}^\star\mathbb{R}$ to which the function c_0 belongs, where for each i in X, $c_0(i) = 0$. Let α in ${}^\star\mathbb{R}$ be such that $f \in \alpha$. Then

$$\{i | c_0(i) + f(i) = f(i)\} = X \in \mathcal{U},$$

and thus $\alpha_0 + \alpha = \alpha$, that is, Statement 6, additive identity element, is true. Statements 1 to 5 and 7 and 8 follow by similar arguments. For example, to show Statement 1, additive commutativity, let β and γ be arbitrary elements of ${}^\star\mathbb{R}$, $g \in \beta$, and $h \in \gamma$, and observe that

$$\{i | g(i) + h(i) = h(i) + g(i)\} = X \in \mathcal{U}. \quad \square$$

Lemma 3.2 *Suppose $\mathcal{U}$ is an ultrafilter on X and $\langle {}^\star\mathbb{R}, \leq, +, \cdot\rangle$ is the $\mathcal{U}$-ultrapower of $\langle \mathbb{R}, \leq, +, \cdot\rangle$. Then $\leq$ is a partial ordering (Definition 1.3) on ${}^\star\mathbb{R}$.*

Proof. (In this proof only filter properties of $\mathcal{U}$ are needed.) It is easy to show that $\leq$ is a reflexive relation on ${}^\star\mathbb{R}$. To show transitivity, assume that α, β, and γ are in ${}^\star\mathbb{R}$ and $\alpha \leq \beta$ and $\beta \leq \gamma$. Let f, g, and h be such that $f \in \alpha$, $g \in \beta$, and $h \in \gamma$. Let

$$A = \{i | f(i) \leq g(i)\} \text{ and } B = \{i | g(i) \leq h(i)\}.$$

Because $\alpha \leq \beta$ and $\beta \leq \gamma$, it follows that $A \in \mathcal{U}$ and $B \in \mathcal{U}$. Because $\leq$ is a transitive relation on $\mathbb{R}$,

$$\{i | f(i) \leq h(i)\} \supseteq (A \cap B) \in \mathcal{U}.$$

Thus $\alpha \leq \gamma$. □

Lemma 3.3 *Suppose $\mathcal{U}$ is an ultrafilter on X and $\langle {}^\star\mathbb{R}, \leq, +, \cdot\rangle$ is the $\mathcal{U}$-ultrapower of $\langle \mathbb{R}, \leq, +, \cdot\rangle$. Then the following two statements are true for all α, β, γ, ζ in ${}^\star\mathbb{R}$:*

1. (monotonic additivity) *If $\alpha \leq \beta$ and $\gamma \leq \zeta$, then $\alpha + \gamma \leq \beta + \zeta$.*
2. (monotonic multiplicativeness) *If $0 < \alpha$ and $0 < \beta$, then $0 < \alpha \cdot \beta$.*

Proof. (In this proof only filter properties of $\mathcal{U}$ are needed.) 1. Suppose α, β, γ, and ζ are arbitrary elements of ${}^\star\mathbb{R}$, $f \in \alpha$, $g \in \beta$, $h \in \gamma$, and $e \in \zeta$, $\alpha \leq \beta$, and $\gamma \leq \zeta$. Let

$$A = \{i | f(i) \leq g(i)\} \text{ and } B = \{i | h(i) \leq e(i)\}.$$

Then A and B are in $\mathcal{U}$, and

$$A \cap B \subseteq \{i | f(i) + h(i) \leq g(i) + e(i)\}.$$

Because $(A \cap B) \in \mathcal{U}$, $\{i | f(i) + h(i) \leq g(i) + e(i)\} \in \mathcal{U}$. Thus

$$\alpha + \gamma \leq \beta + \zeta,$$

and Statement 1 has been shown. Statement 2 follows by a similar argument. □

Using the fact that $\mathcal{U}$ is an ultrafilter—and not just a filter—yields the stronger result below that $\langle {}^\star\mathbb{R}, \leq, +, \cdot\rangle$ is a "totally ordered field."

Definition 3.13 $\mathfrak{A} = \langle A, \preceq, \oplus, \odot, i_0, i_1 \rangle$ is said to be a *totally ordered field (with additive operation* $\oplus$*, multiplicative operation* $\odot$*, additive identity* i_0*, and multiplicative identity* i_1*)* if and only if $\preceq$ is a binary relation on A, $\oplus$ and $\odot$ are binary operations on A, i_0 and i_1 are elements of A, and the following three conditions hold:

(*i*) *Total ordering:* $\preceq$ is a total ordering on A (Definition 1.3).

(*ii*) With appropriate substitutions of A for ${}^\star\mathcal{R}$, $\preceq$ for $\leq$, $\oplus$ for $+$, $\odot$ for $\cdot$, i_0 for c_0 ($= 0$), and i_1 for c_1 ($=1$), the follows conditions of Lemmas 3.1 and 3.3 hold:
(*a*) additive commutativity of $\oplus$
(*b*) multiplicative commutativity of $\odot$
(*c*) additive associativity of $\oplus$
(*d*) multiplicative associativity of $\odot$
(*e*) distributivity of $\odot$ over $\oplus$
(*f*) i_0 is the additive identity element
(*g*) additive inverse
(*h*) i_1 is the multiplicative identity element
(*i*) monotonic additivity
(*j*) monotonic multiplicativeness.

(*iii*) *(multiplicative inverse)* For each x in A, if $x \neq i_0$, then there exists y in A such that $x \odot y = i_1$. □

The real and rational numbers form totally ordered fields with their usual orderings, addition and multiplication operations, and with additive and multiplicative identities respectively, 0 and 1. The following theorem characterizes the totally ordered field of real numbers.

Theorem 3.4 *All Dedekind complete, totally ordered fields are isomorphic to the totally ordered field of real numbers.*

The proof of Theorem 3.4 is well-known in algebra, and the construction of the isomorphism is a straightforward matter. □

Theorem 3.5 *Suppose* $\mathcal{U}$ *is an ultrafilter on* X *and* $\langle {}^\star\mathbb{R}, \leq, +, \cdot \rangle$ *is the* $\mathcal{U}$*-ultrapower of* $\langle \mathbb{R}, \leq, +, \cdot \rangle$*. Let* α_0 *and* α_1 *be the* $\sim$*-equivalence classes such that* $f \in \alpha_0$ *and* $g \in \alpha_1$*, where for all* i *in* X*,*

$$f(i) = 0 \text{ and } g(i) = 1\,.$$

Then $\langle {}^\star\mathbb{R}, \leq, +, \cdot, \alpha_0, \alpha_1 \rangle$ *is a totally ordered field.*

Proof. (This proof uses ultrafilter properties of $\mathcal{U}$.) It is immediate that α_0 is an additive identity and α_1 is a multiplicative identity. Thus, to show that $\langle {}^\star\mathbb{R}, \leq, +, \cdot, \alpha_0, \alpha_1\rangle$ is a totally ordered field, it follows from Lemmas 3.1, 3.2, and 3.3 that it is sufficient to show that $\leq$ is a connected relation on ${}^\star\mathbb{R}$ and that each element to ${}^\star\mathbb{R}$ that is different from the additive identity has a multiplicative inverse.

Let α, β, and γ be arbitrary elements of ${}^\star\mathbb{R}$, and let $f \in \alpha$, $g \in \beta$, and $h \in \gamma$.

Suppose it is not the case that $\alpha \leq \beta$. Therefore, to show $\leq$ is a connected relation on ${}^\star\mathbb{R}$, it needs only be shown that $\beta < \alpha$. Let

$$A = \{i | f(i) \leq g(i)\} .$$

Because it is not the case that $\alpha \leq \beta$, $A \notin \mathcal{U}$. Because $\mathcal{U}$ is an ultrafilter, it then follows from Theorem 3.3 that $X - A$ is in $\mathcal{U}$. However, because $\leq$ is a connected relation on $\mathbb{R}$, it follows that

$$X - A = \{i | g(i) < f(i)\} .$$

Thus, because $X - A$ is in $\mathcal{U}$, it then follows that $\beta < \alpha$, establishing that $\leq$ is a connected relation of ${}^\star\mathbb{R}$.

Suppose $\gamma \neq \alpha_0$. Let

$$B = \{i | h(i) = 0\} .$$

Because $\gamma \neq \alpha_0$, $B \notin \mathcal{U}$. Because $\mathcal{U}$ is an ultrafilter,

$$X - B = \{i | h(i) \neq 0\} \in \mathcal{U} .$$

Let e be the function on X such that for all i in X,

$$e(i) = \begin{cases} h(i)^{-1} & \text{if } h(i) \neq 0 \\ 0 & \text{if } h(i) = 0 . \end{cases}$$

Then,

$$\{i | h(i) \cdot e(i) = 1\} \supseteq (X - B) \in \mathcal{U} .$$

Let ζ be the element of ${}^\star\mathbb{R}$ such that $e \in \zeta$. Because $(X - B) \in \mathcal{U}$, it follows that $\gamma \cdot \zeta = \alpha_1$. $\square$

As illustrated by Theorem 3.5, many true statements about the structure $\mathfrak{R} = \langle \mathbb{R}, \leq, +, \cdot \rangle$ remain true about a $\mathcal{U}$-ultrapower of $\mathfrak{R}$. Theorem 11.3 of Chapter 11 characterizes a large class of statements about $\mathfrak{R}$ that remain true about a $\mathcal{U}$-ultrapower of $\mathfrak{R}$.

3.4 Totally Ordered Field Extensions of the Reals

Definition 3.14 $\mathfrak{A} = \langle A, \preceq, \oplus, \odot, i_0, i_1 \rangle$ is said to be a *totally ordered field extension of the reals* if and only if

(i) $\mathfrak{A}$ is a totally ordered field;

(ii) $\mathbb{R} \subseteq A$, $\leq \subseteq \preceq$, $+ \subseteq \oplus$, and $\cdot \subseteq \odot$, where $+$ and $\cdot$ are respectively the operations of addition and multiplication on the reals; and

(iii) $i_0 = 0$ and $i_1 = 1$.

$\mathfrak{A} = \langle A, \preceq, \oplus, \odot, i_0, i_1 \rangle$ is said to be a *proper* totally ordered field extension of the reals if and only if $\mathfrak{A}$ is a totally odered field extension of the reals and $\mathbb{R} \subset A$. □

Convention 3.4 Let $\mathcal{U}$ be an ultrafilter on X and ${}^{\star}\mathfrak{R} = \langle {}^{\star}\mathbb{R}, \leq, +, \cdot \rangle$ be the $\mathcal{U}$-ultrapower of $\mathfrak{R} = \langle \mathbb{R}, \leq, +, \cdot \rangle$. For each r in $\mathbb{R}$ let c_r be the function on X such that for all i in X, $c_r(i) = r$, and let α_r be the element of ${}^{\star}\mathbb{R}$ such that $c_r \in \alpha_r$. Let φ be the function on $\mathbb{R}$ such that for each r in $\mathbb{R}$,

$$\varphi(r) = \alpha_r .$$

Then φ is an isomorphic imbedding of $\mathfrak{R}$ into ${}^{\star}\mathfrak{R}$. Thus we may consider ${}^{\star}\mathfrak{R}$ to be totally ordered field extension of $\mathfrak{R}$, and we will do so throughout most of the book.

Throughout the book, the usual conventions, notation, and definitions involving the algebra of the real number system are used freely and usually without comment. □

Theorem 3.6 *Suppose $\mathcal{U}$ is an ultrafilter on X and ${}^{\star}\mathfrak{R} = \langle {}^{\star}\mathbb{R}, \leq, +, \cdot \rangle$ is the $\mathcal{U}$-ultrapower of $\mathfrak{R} = \langle \mathbb{R}, \leq, +, \cdot \rangle$. Then the following three statements are true:*

1. *${}^{\star}\mathfrak{R}$ is a totally ordered field extension of the reals.*

2. *If $\mathcal{U}$ is a principal ultrafilter, then ${}^{\star}\mathbb{R} = \mathbb{R}$, that is, ${}^{\star}\mathfrak{R}$ is not a proper ordered field extension of the reals.*

3. *If $X = \mathbb{I}^+$ and $\mathcal{U}$ contains the co-finite subsets of X, then ${}^{\star}\mathfrak{R}$ is a proper ordered field extension of the reals.*

Proof. Statement 1 is immediate from Theorem 3.5. The proof of Statement 2 is direct and left to the reader. To show Statement 3, let f be the function from $\mathbb{I}^+$ such that $f(i) = i$ for each i in $\mathbb{I}^+$, and let α in ${}^{\star}\mathbb{R}$ be such that $f \in \alpha$. Then $\alpha \in ({}^{\star}\mathbb{R} - \mathbb{R})$. □

Definition 3.15 Let $\langle {}^\star\mathbb{R}, \leq, +, \cdot, 0, 1\rangle$ be a totally ordered field extension of the reals, and let $\alpha \in {}^\star\mathbb{R}$. Then α is said to be:

- *Infinitesimal* if and only if $|\alpha| < r$ for each r in $\mathbb{R}^+$.
- *Finite* if and only if $|\alpha| < r$ for some r in $\mathbb{R}^+$.
- *Infinite* if and only if $|\alpha| > r$ for each r in $\mathbb{R}^+$. □

Let ${}^\star\mathfrak{R}$ be a totally ordered field extension of the reals. Then it immediately follows from Definition 3.15 that 0 is infinitesimal and s is finite for each s in $\mathbb{R}$.

Theorem 3.7 *Let* ${}^\star\mathfrak{R} = \langle {}^\star\mathbb{R}, \leq, +, \cdot, 0, 1\rangle$ *be a proper totally ordered field extension of the reals. Then there exist positive infinitesimal and positive infinite elements of* ${}^\star\mathbb{R}$.

Proof. Because ${}^\star\mathfrak{R}$ is a proper extension, let α be an element of ${}^\star\mathbb{R} - \mathbb{R}$. Because $0 \in \mathbb{R}$, $\alpha \neq 0$. Because $\alpha \neq 0$, either $0 < \alpha$ or $\alpha < 0$. If $\alpha < 0$, then $0 < -\alpha$, that is, either $0 < \alpha$ or $0 < -\alpha$. Without loss of generality suppose $0 < \alpha$. There are three cases to consider.

Case 1, $\alpha < r$ for each r in $\mathbb{R}^+$. Then α is a positive infinitesimal. Let r be an arbitrary element of $\mathbb{R}^+$. Because $\alpha < r$ and ${}^\star\mathfrak{R}$ is a totally ordered field, $r^{-1} < \alpha^{-1}$. Because r is an arbitrary of $\mathbb{R}^+$, it follows that for each s in $\mathbb{R}^+$, $s < \alpha^{-1}$. Thus α^{-1} is a positive infinite element of $\mathbb{R}$.

Case 2, $\alpha > r$ for each r in $\mathbb{R}^+$. Then α is a positive infinite element of $\mathbb{R}$. Let r be an arbitrary element of $\mathbb{R}^+$. Because $r < \alpha$ and ${}^\star\mathfrak{R}$ is a totally ordered field, $0 < \alpha^{-1} < r^{-1}$. Because r is an arbitrary of $\mathbb{R}^+$, it follows that for each s in $\mathbb{R}^+$, $\alpha^{-1} < s$. Thus α^{-1} is a positive infinitesimal.

Case 3, there exist r and s in $\mathbb{R}^+$ such that $r < \alpha < s$. Let

$$A = \{t | t \in \mathbb{R}^+ \text{ and } t < \alpha\} \text{ and } B = \{t | t \in \mathbb{R}^+ \text{ and } \alpha \leq t\}.$$

Then (A, B) is a Dedekind cut (Definition 1.5) of $\langle \mathbb{R}^+, \leq\rangle$. Because $\langle \mathbb{R}^+, \leq\rangle$ is Dedekind complete, by Theorem 1.1 let c in $\mathbb{R}^+$ be the cut element of (A, B). Because $\alpha \notin \mathbb{R}^+$, $\alpha \neq c$. Therefore, either $\alpha - c > 0$ or $c - \alpha > 0$. Without loss of generality, suppose $c - \alpha > 0$. (The case $\alpha - c > 0$ follows by a similar argument.) Let r be an arbitrary element of $\mathbb{R}^+$. It will be shown that $c - \alpha < r$, thus establishing that $c - \alpha$ is a positive infinitesimal. Suppose $r \leq c - \alpha$. A contradiction will be shown. Then $r + \alpha \leq c$, and thus

$$\alpha < c - \frac{r}{2}.$$

Let

$$d = c - \frac{r}{2}. \tag{3.1}$$

Then, because $\alpha < d$, d must be in B. Therefore, because c is the cut element of (A, B), $c \leq d$, which contradicts Equation 3.1, because $r > 0$. □

Definition 3.16 Let $\langle {}^\star\mathbb{R}, \leq, +, \cdot, 0, 1\rangle$ be a totally ordered field extension of the reals and α be a finite element of ${}^\star\mathbb{R}$. Let

$$A = \{t | t \in \mathbb{R}^+ \text{ and } t < \alpha\} \text{ and } B = \{t | t \in \mathbb{R}^+ \text{ and } \alpha \leq t\}.$$

Then, because α is finite, (A, B) is a Dedekind cut of $\langle \mathbb{R}^+, \leq\rangle$. Let s be the cut element of (A, B). Then, by definition, ${}^\circ\alpha = s$. ${}^\circ\alpha$ is called the *standard part* of α. □

Theorem 3.8 *Let $\langle {}^\star\mathbb{R}, \leq, +, \cdot, 0, 1\rangle$ be a proper totally ordered field extension of the reals and α be a finite element of ${}^\star\mathbb{R}$. Then $|{}^\circ\alpha - \alpha|$ is infinitesimal.*

Proof. Let r be an arbitrary element of $\mathbb{R}^+$ and

$$A = \{x \,|\, x \in \mathbb{R}^+ \text{ and } x \leq \alpha\} \text{ and } B = \{x \,|\, x \in \mathbb{R}^+ \text{ and } x > \alpha\}.$$

Because ${}^\circ\alpha$ is the cut element of the Dedekind cut (A, B),

$$({}^\circ\alpha - \frac{r}{2}) \in A \text{ and } ({}^\circ\alpha + \frac{r}{2}) \in B.$$

But then,

$${}^\circ\alpha - \frac{r}{2} \leq \alpha \leq {}^\circ\alpha + \frac{r}{2}.$$

In other words,

$$|{}^\circ\alpha - \alpha| \leq \frac{r}{2}.$$

Because r is an arbitrary element of $\mathbb{R}^+$, it then follows from Definition 3.15 that $|{}^\circ\alpha - \alpha|$ is infinitesimal. □

Theorem 3.9 *Let $\langle {}^\star\mathbb{R}, \leq, +, \cdot, 0, 1\rangle$ be a proper totally ordered field extension of the reals and α and β be finite elements of $\mathbb{R}$ such that $\alpha < \beta$. Then ${}^\circ\alpha \leq^\circ \beta$.*

Proof. Because for each r in $\mathbb{R}^+$,

$$|{}^\circ\alpha - \alpha| < r \text{ and } |{}^\circ\beta - \beta| < r,$$

it follows that for each r in $\mathbb{R}^+$,

$${}^\circ\beta - {}^\circ\alpha + 2r = ({}^\circ\beta + r) + (-{}^\circ\alpha + r) > \beta - \alpha > 0.$$

That is, for each r in $\mathbb{R}^+$,

$${}^\circ\beta - {}^\circ\alpha + 2r > 0.$$

Thus ${}^\circ\beta - {}^\circ\alpha \geq 0$. Therefore ${}^\circ\alpha \leq {}^\circ\beta$. □

Theorem 3.10 *Let $\langle {}^\star\mathbb{R}, \leq, +, \cdot, 0, 1\rangle$ be a proper totally ordered field extension of the reals and α and β be finite elements of ${}^\star\mathbb{R}$. Then $\alpha+\beta$ is a finite element of ${}^\star\mathbb{R}$ and ${}^\circ(\alpha+\beta) = {}^\circ\alpha + {}^\circ\beta$.*

Proof. It is immediate from Definition 3.15 that $\alpha+\beta$ is finite. By Theorem 3.8, for each r in $\mathbb{R}^+$,

$$|{}^\circ\alpha - \alpha| < r\,, \ |{}^\circ\beta - \beta| < r\,, \text{ and } |{}^\circ(\alpha+\beta) - (\alpha+\beta)| < r\,.$$

Thus, for each r in ${}^\star\mathbb{R}$,

$$|{}^\circ(\alpha+\beta) - ({}^\circ\alpha + {}^\circ\beta)| < 3r\,.$$

Therefore, ${}^\circ(\alpha+\beta) = {}^\circ\alpha + {}^\circ\beta$. □

Theorem 3.11 *Let $\langle {}^\star\mathbb{R}, \leq, +, \cdot, 0, 1\rangle$ be a proper totally ordered field extension of the reals and $\alpha_1, \ldots, \alpha_n$ be finite elements of ${}^\star\mathbb{R}$. Then $\alpha_1 + \cdots + \alpha_n$ is a finite element of ${}^\star\mathbb{R}$ and*

$${}^\circ(\alpha_1 + \cdots + \alpha_n) = {}^\circ\alpha_1 + \cdots + {}^\circ\alpha_n\,.$$

Proof. Left to reader. □

Theorem 3.12 *Let $\langle {}^\star\mathbb{R}, \leq, +, \cdot, 0, 1\rangle$ be a proper totally ordered field extension of the reals, α and α_1 be infinitesimal elements of ${}^\star\mathbb{R}$, β and β_1 be finite elements of ${}^\star\mathbb{R}$, and γ and γ_1 be infinite elements of ${}^\star\mathbb{R}$. Then the following eight statements are true:*

1. *$\alpha + \alpha_1$ is infinitesimal.*
2. *$\alpha \cdot \beta$ is infinitesimal.*
3. *If $\beta \cdot \beta_1$ is not infinitesimal, then $\beta \cdot \beta_1$ is finite and not infinitesimal.*
4. *$\beta + \beta_1$ is finite.*
5. *$\beta + \gamma$ is infinite.*
6. *If β is not infinitesimal, then $\beta \cdot \gamma$ is infinite.*
7. *$\gamma \cdot \gamma_1$ is infinite.*
8. *If β is not infinitesimal, then β^{-1} is finite and not infinitesimal.*

Proof. Left to reader. □

Chapter 4

Qualitative Probability

4.1 Extensive Measurement of Probability

Kolmogorov axioms assume probabilities (numbers) have been assigned to events. The axioms are silent about how this came about. A subarea in the foundations of science called "measurement theory" deals directly with this issue, and more generally with how numbers enter into science and their proper use within science. In the context of probability theory, measurement theory provides a qualitative foundation for the Kolmogorov concepts of "probability" and "probability function" by producing a theory of probability based on qualitative ordering of events in terms of their likelihood of occurrence; that is, measurement theory provides a theory of probability based on an ordering relation $\precsim$, where "$A \precsim B$" is interpreted as, "The event A has a less or equally likely chance of occurring than the event B."

The reduction of scientific quantities to qualitative relationships goes back to at least the late 19th century when H. von Helmholtz (1887) provided a measurement theory for fundamental physical quantities like length, mass, etc. His theory was based on the idea that each fundamental physical quantity had associated with it a natural, observable, *qualitative* ordering and a *qualitative* combination operation (called *concatenation*) that had certain specific, observable, algebraic properties. Hölder (1901) presented a deeper mathematical analysis of Helmholtz's theory of physical measurement.

The qualitative structures developed by Helmholtz for measuring are today called "Dedekind complete extensive structures." An axiomatization of them is provided in the following definition.

Definition 4.1 $\mathfrak{X} = \langle X, \precsim, \oplus \rangle$ is said to be a *Dedekind complete extensive structure* if and only if the following seven axioms hold for all x, y, and z in X:

1. *Total Ordering*: $\precsim$ is a total ordering on X.

2. *Density*: If $x \prec z$ then for some w in X, $x \prec w \prec z$.

3. *Associativity*: $\oplus$ is a binary operation that is *associative*; that is,

$$(x \oplus y) \oplus z = x \oplus (y \oplus z).$$

4. *Monotonicity*:

$$x \precsim y \text{ iff } x \oplus z \precsim y \oplus z \text{ iff } z \oplus x \precsim z \oplus y.$$

5. *Solvability*: If $x \prec y$, then for some w in X, $y = x \oplus w$.

6. *Positivity*: $x \prec x \oplus y$ and $y \prec x \oplus y$.

7. *Dedekind Completeness*: $\langle X, \precsim \rangle$ is Dedekind complete (Definition 1.5). □

In essence Helmholtz (1887) showed the following Theorem.

Theorem 4.1 *Suppose* $\mathfrak{X} = \langle X, \precsim, \oplus \rangle$ *is a Dedekind complete extensive structure and* $\mathfrak{N} = \langle \mathbb{R}^+, \leq, + \rangle$. *Then* (i) *there exists an isomorphism* φ *of* $\mathfrak{X}$ *onto* $\mathfrak{N}$, *and* (ii) *for all isomorphisms* ψ *and* θ *of* $\mathfrak{X}$ *into* $\mathfrak{N}$, *there exists* r *in* $\mathbb{R}^+$ *such that* $\psi = r\theta$. □

Proofs of generalizations of Theorem 4.1 can be found in Chapter 3 of Krantz et al. (1971) and in Section 9 of Chapter 2 of Narens (1985).[1]

An example of extensive measurement is the measurement of mass by an equal arm pan balance. Physical objects a and b are said to be "equivalent in mass," $a \sim b$, if and only if when placed in opposite pans, a balances b. It is assumed that $\sim$ is an equivalence relation. Let X be the set of $\sim$-equivalence classes of physical objects. The binary relation $\precsim$ is defined on X as follows:

> $\alpha \precsim \beta$ if and only if there exist physical objects x in α and y in β such that if x and y are placed in opposite pans, either they balance, or the pan with y becomes lower than the one with x.

[1]To obtain a proof of Theorem 4.1 from these generalizations, one merely uses solvability and Dedekind completeness to show that $\mathfrak{X}$ satisfies the "Archimedean axiom" and simple consequences of the axioms to show that all the isomorphisms are onto $\mathbb{R}^+$.

$\oplus$ is defined on X as follows:

> $\alpha \oplus \beta = \gamma$ if and only if there exist x in α, y in β, and z in γ such that when x and y are placed in the same pan and z in the opposite, the result balances.

It is assumed that $\langle X, \preceq, \oplus \rangle$ satisfies the axioms of a Dedekind complete extensive structure (Definition 4.1).

Another example is the measurement of length. Here R is the set of measuring rods, which ideally look like line segments. Rods a and b are said to be "equivalent in length," $a \sim b$, if and only if a and b can be laid side by side with endpoints exactly corresponding. It is assumed that $\sim$ is an equivalence relation. Let X be the set of $\sim$-equivalence classes of elements of R. Then $\preceq$ is defined on X as follows:

> $\alpha \preceq \beta$ if and only if there exist x in α and y in β such that either $x \sim y$ or when x and y are placed side by side with left endpoints exactly corresponding, then the right endpoint of y extends beyond the right endpoint of x.

$\oplus$ is defined on X as follows:

> $\alpha \oplus \beta = \gamma$ if and only if there exist x in α, y in β, and z in γ such that when x and y are placed on an oriented line with the right endpoint of x touching the left endpoint of y (i.e., "x is abutted to y"), they form a rod w such that $w \sim z$.

In classical theoretical physics, it is assumed that $\langle X, \preceq, \oplus \rangle$ satisfies the axioms of a Dedekind complete extensive structure. However, in practice, mass and length are no longer measured as described above. Instead, they are measured using rather sophisticated instrumentation, and the justifications for the correctness of the resulting measurements rest heavily on physical theory, which in turn assumes at the theoretical level a theory of measurement for mass and length, for example, a theory of measurement based on Dedekind complete extensive structures. Therefore, extensive measurement, as described above, is still useful as a theory of measurement for *theoretical* physics, because it justifies how numbers are assigned to ideal physical entities. Many believe it is important in physical theory to make explicit the measurement process, because the rules by which numbers are assigned to physical entities necessarily condition the mathematical form of physical laws.

Luce (1967) saw that the measurement of uncertainty—that is, the assigning of numerical probabilities to uncertain events—can be looked at as a form of extensive measurement. Let $A \uplus B$ stand for the union of the disjoint events A and B. Then qualitatively, $\uplus$ is the equivalent of adding probabilities. However, unlike in other measurement situations where one must verify either empirically or by other means the associativity of the qualitative concatenation operation, it is automatic for uncertainty measurement that $\uplus$ is an associative operation. Thus one of the key conditions of extensive measurement—associativity—is automatically met. Suppose $A \precsim B$ stands for "A is less or equally likely than B." Then another key condition of extensive measurement—monotonicity—is expressed as follows for uncertainty measurement:

Definition 4.2 Let $\langle X, \mathcal{E} \rangle$ be an algebra of events and $\precsim$ be a binary relation on $\mathcal{E}$. Then $\langle X, \mathcal{E}, \precsim \rangle$ is said to satisfy $\cup$-*monotonicity* if and only if for all A, B, and C in $\mathcal{E}$, if $C \cap A = C \cap B = \varnothing$, then

$$A \precsim B \text{ iff } A \cup C \precsim B \cup C \,. \quad \square$$

Of course, there are differences between the measuring of uncertainty and the measuring of fundamental physical qualities. The most important is that in uncertainty measurement there exists a largest event—the sure event—while in classical extensive measurement, the to-be-measured objects are unbounded in size. Changing the axioms of extensive measurement so that they apply to bounded situations is conceptually a straightforward matter; however, certain technical difficulties arise in amending the method of proof. Luce and Marley (1969) and Krantz, et al. (1971) developed new proof techniques for the bounded case.

Definition 4.3 $\oplus$ is said to be a *partial (binary) operation* on X if and only if $\oplus$ is function on $A \times B$ (called its *domain*) into X, where A and B are some nonempty subsets of X. $x \oplus y$ is said to be *defined* if and only if (x, y) is in the domain of $\oplus$. $\square$

Definition 4.4 $\langle X, \precsim, \oplus \rangle$ is said to be an *extensive structure with maximal element* m if and only if $\precsim$ is a binary relation on X, $\oplus$ is a partial binary operation on X, and m is an element of X such that the following seven axioms are satisfied for all x, y, and z in X:

1. $\precsim$ is a weak ordering (Definition 1.3) on X.

2. If $x \oplus y$ is defined and $u \precsim x$ and $v \precsim y$, then $u \oplus v$ is defined.

3. If $x \oplus y$ is defined, then $x \prec x \oplus y$.

4. If $x \oplus (y \oplus z)$ is defined, then $(x \oplus y) \oplus z$ is defined and

$$x \oplus (y \oplus z) \sim (x \oplus y) \oplus z\,.$$

5. If $x \prec y$, then there exists w in X such that $x \oplus w \precsim y$.

6. There exist u and v in X such that $u \oplus v = m$.

7. *(Archimedean Axiom)* There does not exist an infinite sequence of elements of X, $x_1, \ldots, x_i, \ldots$, such that for each positive integer j, $x_{j+1} = x_j \oplus x_1$. □

Theorem 4.2 *Let $\langle X, \precsim, \oplus\rangle$ be an extensive structure with maximal element m. Then the following two statements are true:*

1. (Existence) *There exists a function φ from X into $\mathbb{R}^+$ such that for all x and y in X,*

$$x \precsim y \;\; \textit{iff} \;\; \varphi(x) \leq \varphi(y)\,, \tag{4.1}$$

and, if $x \oplus y$ is defined then

$$\varphi(x \oplus y) = \varphi(x) + \varphi(y)\,. \tag{4.2}$$

2. (Uniqueness) *If φ and ψ satisfy Equations 4.1 and 4.2, then (*i*) there exists r in $\mathbb{R}^+$ such that $\varphi = r\psi$, and (*ii*) for each s in $\mathbb{R}^+$, $s\varphi$ satisfies Equations 4.1 and 4.2.*

Proof. Theorem 3 of Section 3.5 of Krantz, et al. (1971) and Statement 6 of Definition 4.4. □

Let $\langle X, \precsim, \oplus\rangle$ be an extensive structure with maximal element m. Then it is an immediate consequence of Theorem 4.2 that there is exactly one function φ satisfying Equations 4.1 and 4.2 such that $\varphi(m) = 1$.

Extensive structures with maximal elements occur in the measurement of uncertainty as follows:

Suppose $\mathcal{E}$ is a boolean algebra of subsets of X and $\precsim$ is a weak ordering on $\mathcal{E}$. For A and B in $\mathcal{E}$, interpret $A \precsim B$ as "the event A has less or equally likely chance of occurring than the event B." Let $\mathcal{E}'$ be the set of $\sim$-equivalence classes of $\mathcal{E}$, and define $\precsim'$ on $\mathcal{E}'$ as follows: for all Γ and Σ in $\mathcal{E}'$, $\Gamma \precsim' \Sigma$ if and only if there exist B in Γ and C in Σ such that $B \precsim C$. Let $\mathcal{E}''$ be the set that consists of those equivalence classes in $\mathcal{E}'$ that have elements B such that $\varnothing \prec B$. Intuitively, $\mathcal{E}''$ consists of equivalence classes of events that have "positive likelihood." Let $\precsim''$ be the restriction of $\precsim'$ to $\mathcal{E}''$. Define $\oplus$ on $\mathcal{E}''$ as follows: For all Γ, Δ, and Σ in $\mathcal{E}''$,

$$\Gamma \oplus \Delta = \Sigma \text{ if and only if}$$
$$\text{for some } A \in \Gamma \text{ and } B \in \Delta \text{ such that } A \cap B = \varnothing,\ (A \cup B) \in \Sigma\,.$$

Then if $\langle X, \mathcal{E}, \precsim \rangle$ satisfies certain plausible conditions (which hold in the most important probability structures used in mathematics), then $\langle \mathcal{E}'', \precsim'', \oplus \rangle$ is an extensive structure with a maximal element. It then easily follows from Theorem 4.2 that there is a unique finitely additive probability function $\mathbb{P}$ on $\mathcal{E}$ such that for all A and B in $\mathcal{E}$,

$$A \precsim B \text{ iff } \mathbb{P}(A) \le \mathbb{P}(B)\,.$$

The following theorem of Luce (1967) provides an example of such "plausible conditions:"

Theorem 4.3 *Suppose X is a nonempty set, $\mathcal{E}$ is an algebra of subsets of X, and $\precsim$ is a binary relation on $\mathcal{E}$ such that the following five statements hold for all A, B, C, and D in $\mathcal{E}$:*

1. *$\precsim$ is a weak ordering on $\mathcal{E}$.*
2. *$\varnothing \precsim A \precsim X$ and $\varnothing \prec X$.*
3. *If $A \cap B = A \cap C = \varnothing$, then*
$$B \precsim C \text{ iff } A \cup B \precsim B \cup C\,.$$
4. *If $A \cap B = \varnothing$, $C \prec A$, and $D \precsim B$, then there exists C', D', and E in $\mathcal{E}$ such that*
   - (i) *$E \sim A \cup B$;*
   - (ii) *$C' \cap D' = \varnothing$;*
   - (iii) *$C' \cup D' \subset E$; and*
   - (iv) *$C' \sim C$ and $D' \sim D$.*
5. (Archimedean Axiom) *There do not exist an element A in $\mathcal{E}$ and infinite sequences of elements of $\mathcal{E}$, $A_1, \ldots, A_i, \ldots,$ $B_1, \ldots, B_i, \ldots,$ and $C_1, \ldots, C_i, \ldots$ such that*
   - (i) *$A_1 = B_1$ and $B_1 \sim A$;*
   - (ii) *$B_i \cap C_i = \varnothing$;*

(iii) $B_i \sim A_i$;

(iv) $C_i \sim A$; *and*

(v) $A_{i+1} = B_i \cup C_i$.

Then there exists a unique finitely additive probability function $\mathbb{P}$ *on* $\mathcal{E}$ *such that for all* F *and* G *in* $\mathcal{E}$,

$$F \precsim G \text{ iff } \mathbb{P}(F) \leq \mathbb{P}(G).$$

Proof. See Luce (1967), or pp. 95–97 of Narens, (1985), or pp. 211–214 of Krantz et al., (1971) for details. □

Krantz, et al. (1971) makes the following observations about the measurement of uncertainty:

> Theories about the representation of a qualitative ordering of events have generally been classed as subjective (de Finetti, 1937), intuitive (Koopman, 1940 a,b; 1941) , or personal (Savage, 1954), with the intent of emphasizing that the ordering relation $\precsim$ *may* be peculiar to an individual and that he *may* determine it by any means at his disposal, including his personal judgment. But these "mays" in no way preclude orderings that are determined by well-defined, public, and scientifically agreed upon procedures, such as counting relative frequencies under well-specified conditions. Even these objective procedures often contain elements of personal judgment; for example, counting the relative frequency of heads depends on our judgment that the events "heads on trial 1" and "heads on trial 2" are equiprobable. This equiprobability judgment is part of a partly "objective," partly "subjective" ordering of events.
>
> Presumably, as science progresses, objective procedures will come to be developed in domains for which we have little alternative but to accept the considered judgments of informed and experienced individuals.
>
> Ellis pointed out that the development of a probability ordering is
>
> > ... analogous to that of finding a thermometric property, which... was the first step towards devising a temperature scale.
> >
> > The comparison between probability and temperature may be illuminating in other ways. The first

> thermometers were useful mainly for comparing atmospheric temperatures. The air thermometers of the 17th century, for example, were not adaptable for comparing or measuring temperatures of small solid objects. Consequently, in the early history of thermometry, there were many things which possessed temperature which could not be fitted into an objective temperature order. Similarly, then, we should not necessarily expect to find any single objective procedure capable of ordering all propositions in respect of probability, even if we assume that all propositions possess probability. Rather, we should expect there to be certain kinds of propositions that are much easier to fit into an objective probability order than others... (Ellis, 1966, p. 172).

Since virtually all representation theorems yield a unique probability measure, agreement about the probability ordering of an algebra of events or, as is more usual, agreement about a method to determine that ordering is sufficient to defined a unique numerical probability—an objective probability *(pp. 201–202).*

4.2 Probability Representations

Definition 4.5 Suppose that $\mathcal{E}$ is a boolean algebra of subsets of X and $\precsim$ is a reflexive relation on $\mathcal{E}$.

$\mathbb{P}$ is said to be a *probability representation* of $\langle X, \mathcal{E}, \precsim \rangle$ if and only if $\mathbb{P}$ is a function from $\mathcal{E}$ into the closed interval $[0,1]$ of $\langle \mathbb{R}, \leq \rangle$ such that the following four statements hold for all A and B in $\mathcal{E}$:

1. $\mathbb{P}(\varnothing) = 0$ and $\mathbb{P}(X) = 1$.
2. If $A \prec B$, then $\mathbb{P}(A) < \mathbb{P}(B)$.
3. if $A \sim B$, then $\mathbb{P}(A) = \mathbb{P}(B)$.
4. If $A \cap B = \varnothing$, then $\mathbb{P}(A \cup B) = \mathbb{P}(A) + \mathbb{P}(B)$. □

Because in Definition 4.5 $\precsim$ is only a reflexive ordering and not a weak ordering, Statement 2 has the form "if ... then" rather than "... iff ...".

In algebra if one is presented with left and right columns of numbers and ask to find which side has, if either, the greatest sum, then one can use cancellation to simplify the problem. That is, if a number occurs in both and the right columns than one can eliminate (cancel) one occurrence of it in the left column and one occurrence of it in the right column and then compare the sums of the altered columns. They will have the same ordering as the sums of the original columns. This process can be repeated on the altered columns.

Let $\#_l(x)$ stand for the number of occurrences of the number x in the left column and $\#_r(x)$ the number of its occurrences in the right column. (If x does not occur in the left column then $\#_l(x) = 0$; similarly $\#_r(x) = 0$ stands for x not occurring in the right column.) In the special circumstance in which $\#_l(y) = \#_r(y)$ for all numbers y, it must be the case that the sums of the left and right columns are the same. Note that this conclusion is reached through repeated cancellations. The sums of the left and right columns do not need to be computed. Surprisingly, a related form of cancellation provides necessary and sufficient qualitative conditions for the existence of a probability representation of structures of the form $\langle X, \mathcal{E}, \precsim \rangle$, where $\mathcal{E}$ is a finite boolean algebra.

Definition 4.6 Suppose that $\mathcal{E}$ is a boolean algebra of subsets of a finite set X, $\precsim$ is a reflexive relation on $\mathcal{E}$, and $\mathfrak{X} = \langle X, \mathcal{E}, \precsim \rangle$.

Let Γ be a finite set of $\precsim$-equivalences of elements of $\mathcal{E}$ or $\precsim$-strict inequalities of elements of $\mathcal{E}$; that is, each γ in Γ has the form $A \sim B$ or the form $A \prec B$, where A and B are elements of $\mathcal{E}$. By convention, Γ is allowed to contain several copies of an equivalence or a strict inequality. (To be precise, we should consider Γ to be an *indexed family* rather than a *set*.)

For each x in X, define $\Gamma_l(x)$ and $\Gamma_r(x)$ as follows:

(*i*) $\Gamma_l(x)$ is the number of γ in Γ such that x is a member of the left side of γ; and

(*ii*) $\Gamma_r(x)$ is the number of γ in Γ such that x is a member of the right side of γ.

For each k in $\mathbb{I}^+$, $\mathfrak{X}$ is said to satisfy the kth *cancellation axiom* if and only if for each Γ which is a set of equivalences or strict inequalities of elements of $\mathcal{E}$, if Γ has cardinality k, and for each x in X,

$$\Gamma_l(x) = \Gamma_r(x)\,,$$

then each member of Γ is an equivalence.

$\mathfrak{X}$ is said to satisfy the *finite cancellation axioms* if and only if $\mathfrak{X}$ satisfies the kth cancellation axiom for each k in $\mathbb{I}^+$. □

Theorem 4.4 *Suppose that X is finite and $\mathfrak{X} = \langle X, \mathcal{E}, \precsim \rangle$ has a probability representation. Then $\mathfrak{X}$ satisfies the finite cancellation axioms.*

Proof. Let $\mathbb{P}$ be a probability representation of $\mathfrak{X}$. Let k be an arbitrary element of $\mathbb{I}^+$, $\Gamma = \{\gamma_1, \ldots, \gamma_k\}$, where for i, $i = 1, \ldots, k$, γ_i is either a $\precsim$-equivalence of elements of $\mathcal{E}$ or a $\precsim$-strict inequality of elements of $\mathcal{E}$. Suppose for each x in X, $\Gamma_l(x) = \Gamma_r(x)$. We need only to show that Γ does not contain a $\precsim$-strict inequality.

Suppose Γ contains a $\precsim$-strict inequality. A contradiction will be shown. For $i = 1, \ldots, k$, let

$$\gamma_i = A_i \lhd_i B_i, \text{ where } \lhd_i \text{ is either } \sim \text{ or } \prec.$$

Then for $i = 1, \ldots, k$,

$$\text{if } \lhd_i \text{ is } \prec, \text{ then } \mathbb{P}(A_i) < \mathbb{P}(B_i),$$

and

$$\text{if } \lhd_i \text{ is } \sim, \text{ then } \mathbb{P}(A_i) = \mathbb{P}(B_i).$$

Therefore, because by hypothesis Γ contains a $\precsim$-strict inequality,

$$\sum_{i=1}^{k} \mathbb{P}(A_i) < \sum_{i=1}^{k} \mathbb{P}(B_i).$$

However, because by hypothesis X is finite, it follows that for each C in $\mathcal{E}$,

$$\mathbb{P}(C) = \sum_{x \in C} \mathbb{P}(\{x\}),$$

and thus,

$$\sum_{i=1}^{k} \sum_{x \in A_i} \mathbb{P}(\{x\}) = \sum_{i=1}^{k} \mathbb{P}(A_i) < \sum_{i=1}^{k} \mathbb{P}(B_i) = \sum_{i=1}^{k} \sum_{x \in B_i} \mathbb{P}(\{x\}). \tag{4.3}$$

However, because for each x in X, $\Gamma_l(x) = \Gamma_r(x)$, it follows that

$$\sum_{i=1}^{k} \sum_{x \in A_i} \mathbb{P}(\{x\}) = \sum_{i=1}^{k} \sum_{x \in B_i} \mathbb{P}(\{x\}),$$

which contradicts Equation 4.3. $\square$

4.2.1 Scott's Theorem

Theorem 4.5 (Scott's Theorem) *Suppose X is finite, $\mathfrak{X} = \langle X, \mathcal{E}, \precsim \rangle$, where $\mathcal{E}$ is a boolean algebra of subsets of X and $\precsim$ is a reflexive relation on $\mathcal{E}$ such that $\varnothing \prec X$. Suppose $\mathfrak{X}$ satisfies the finite cancellation axioms. Then $\mathfrak{X}$ has a probability representation.* (Scott, 1964.)

Idea of the proof

Suppose X is finite, $\mathfrak{X} = \langle X, \mathcal{E}, \precsim \rangle$, where $\mathcal{E}$ is a boolean algebra of subsets of X and $\precsim$ is a reflexive relation on $\mathcal{E}$ such that $\varnothing \prec X$, and suppose A and B are elements of $\mathcal{E}$. Let n in $\mathbb{I}^+$ be the number of elements of X, and suppose $\{a_1, \ldots, a_n\}$ is a listing of the elements of X. Then each E in $\mathcal{E}$ can be identified with a vector $(e_1, \ldots, e_n)$ of $\mathbb{R}^n$, where for $1 \leq i \leq n$,

$$e_i = \begin{cases} 1 & \text{if } e_i \in E \\ 0 & \text{if } e_i \notin E\,. \end{cases}$$

Then, for example, "$A \prec B$ and there is a probability representation of $\mathfrak{X}$" implies "The inequality,

$$a_1 x_1 + \ldots + a_n x_n < b_1 x_1 + \ldots + b_n x_n\,,$$

has a solution $x_i = p_i$, where $0 \leq p_i$, $\sum p_i = 1$, and $(a_1, \ldots, a_n)$ and $(b_1, \ldots, b_n)$ are respectively identifications of A and B." In a similar manner a finite set of $\precsim$-inequalities and $\precsim$-equivalences and, "There is a probability representation of $\mathfrak{X}$," implies a simultaneous solution to a system of real inequalities and equivalences appropriately generated by the identifications of the elements of $\mathcal{E}$ occurring in the $\precsim$-strict inequalities and $\precsim$-equivalences. The finite cancellation axioms provide qualitative conditions that allow a standard lemma to be invoked about the simultaneous solution for a system of strict inequalities and equations with real coefficients. The resulting solution yields a probability representation.

A needed lemma

The following well-known lemma concerning the simultaneous solution to a finite number of strict inequalities and equations in n unknowns is used in the proof of Scott's Theorem.

Lemma 4.1 *The system of the following k strict real inequalities in the unknowns $x_1, \ldots, x_n$,*

$$a_1^1 x_1 + \cdots + a_n^1 x_n \quad > \quad 0$$

$$
\begin{aligned}
a_1^2 x_1 + \cdots + a_n^2 x_n &> 0 \\
&\vdots \\
a_1^k x_1 + \cdots + a_n^k x_n &> 0\,,
\end{aligned}
$$

and $m-k$ real equations,

$$
\begin{aligned}
a_1^{k+1} x_1 + \cdots + a_n^{k+1} x_n &= 0 \\
a_1^{k+2} x_1 + \cdots + a_n^{k+2} x_n &= 0 \\
&\vdots \\
a_1^m x_1 + \cdots + a_n^m x_n &= 0\,,
\end{aligned}
$$

either has a simultaneous real solution, $p_1 = x_1$, ..., $p_n = x_n$, or there exist non-negative real numbers $r_1, \ldots, r_k$, not all of which are 0, and real numbers $r_{k+1}, \ldots, r_m$ such that for $j = 1, \ldots, n$,

$$
\sum_{i=1}^{m} r_i a_j^i = 0\,.
$$

Furthermore, if a_j^i is rational for $j = 1, \ldots, n$ and $i = 1, \ldots, m$, then $r_1, \ldots, r_k$ can be chosen to be rational.

Proof. A nice proof and discussion of this well-known lemma is given in Chapter 2 of Krantz et al. □

Proof of Scott's Theorem (Theorem 4.5)

(The reader may want to skip the proof of Theorem 4.5 on first reading.) Let Γ be the set of strict inequalities that hold between elements of $\mathcal{E}$ and Σ be the set of equivalences that hold between elements of $\mathcal{E}$. Because X is finite, $\mathcal{E}$ is finite, and therefore Γ and Σ are finite. Let n be the cardinality of X and $X = \{x_1, \ldots, x_n\}$. For each A in $\mathcal{E}$, let $\overline{A}$ be the following vector in $\mathbb{R}^n$:

$$
\overline{A} = (a_1, \ldots, a_n)\,,
$$

where for $i = 1, \ldots, n$,

$$
a_i = \begin{cases} 1 & \text{if } x_i \in A \\ 0 & \text{if } x_i \notin A\,. \end{cases}
$$

For each γ in $\Gamma \cup \Sigma$, if γ is $A \prec B$ or is $A \sim B$, let

$$\overline{\gamma} = \overline{A} - \overline{B}.$$

It will be shown by contradiction that there exists a vector c in $\mathbb{R}^n$ such that for all γ,

$$\text{if } \gamma \in \Gamma \text{ then } c \cdot \overline{\gamma} > 0 \tag{4.4}$$

and

$$\text{if } \gamma \in \Sigma \text{ then } c \cdot \overline{\gamma} = 0. \tag{4.5}$$

Suppose that there does not exist c in $\mathbb{R}^n$ that satisfy Equations 4.4 and 4.5 for all γ in $\Gamma \cup \Sigma$. Because $\varnothing \prec X$ and $X \sim X$, it follows that $\Gamma \neq \varnothing$ and $\Sigma \neq \varnothing$. Let

$$\Gamma = \{\gamma^1, \ldots, \gamma^k\} \text{ and } \Sigma = \{\gamma^{k+1}, \ldots, \gamma^p\}$$

be listings of the elements of Γ and Σ. Then by Lemma 4.1, let $r_1, \ldots, r_p$ be rational numbers such that

- $r_1, \ldots, r_k$ are nonnegative,
- not all of $r_1, \ldots, r_k$ are 0,
- and

$$\sum_{i=1}^{p} r_i \overline{\gamma}_j^i = 0 \text{ for } j = 1, \ldots, n, \tag{4.6}$$

where of course $\overline{\gamma}_j^i$ is the jth coordinate of the vector $\overline{\gamma}^i$.

In order to make use of the finite cancellation axioms in this proof, we need an analog to Equation 4.6 in which r_i are nonnegative integers. To create such analog, define and s_i and λ^i for $i = 1, \ldots, p$ as follows:

- If $i = 1, \ldots, k$, then $s_i = r_i$ and $\lambda^i = \gamma^i$.
- If $i = k+1, \ldots, p$ and r_i is nonnegative, then $s_i = r_i$ and $\lambda^i = \gamma^i$.
- If $i = k+1, \ldots, p$ and r_i is negative and A and B are such that $\gamma^i = A \sim B$, then $s_i = -r_i$ and $\lambda^i = B \sim A$.

The result of this is that for $i = 1, \ldots, p$, s_i is nonnegative and $s_i \overline{\lambda}_j^i = r_i \overline{\gamma}_j^i$. Thus

$$\sum_{i=1}^{p} s_i \overline{\lambda}_j^i = \sum_{i=1}^{p} r_i \overline{\gamma}_j^i = 0 \quad \text{for } j = 1, \ldots, n.$$

Multiplication of both sides of the equation

$$\sum_{i=1}^{p} s_i \overline{\lambda}_j^i = 0$$

by the common denominator w of the positive rational numbers s_i, $i = 1, \ldots, p$, yields

$$\sum_{i=1}^{p} t_i \overline{\lambda}_j^i = 0 \quad \text{for } j = 1, \ldots, n,$$

where $t_i = ws_i$, $i = 1, \ldots, p$, are nonnegative integers.

Let

$$\Lambda' = \{\lambda^i \,|\, 1 \le i \le p\}.$$

For each λ in Λ' let, by definition, for nonnegative integers m, $m\{\lambda\}$ be an (indexed) set of strict inequalities or equivalences that contains exactly m copies of λ. Let

$$\Lambda = t_1\{\lambda^1\} \cup \cdots \cup t_p\{\lambda^p\}.$$

Then for $j = 1, \ldots, n$,

$$\sum_{i=1}^{p} t_i \overline{\lambda^i}_j = 0, \tag{4.7}$$

where t_i, $i = 1, \ldots, p$, are nonnegative integers. Let x_j be an arbitrary element of X. Then it follows from Equation 4.7 that the number of times x_j occurs on the left side of an element of Λ of the form λ^i is the same as the number of times it occurs on the right side of λ^i. Thus for each x in X,

$$\Lambda_l(x) = \Lambda_r(x).$$

Thus, because Λ is a finite set and $\mathfrak{X}$ satisfies the finite cancellation axioms, Λ is a set of equivalences. However, because $\varnothing \prec X$ is Γ and therefore is in Λ, Λ contains a strict inequality, which contradicts that Λ is a set equivalences.

The just previous argument by contradiction shows that Equations 4.4 and 4.5 hold. Thus let c be a vector of $\mathbb{R}^n$ such that for all γ in $\Gamma \cup \Sigma$, Equations 4.4 and 4.5 hold. Then it follows that for each A and B in $\mathcal{E}$,

$$\text{if } A \prec B \text{ then } c \cdot \overline{A} < c \cdot \overline{B} \tag{4.8}$$

and

$$\text{if } A \sim B \text{ then } c \cdot \overline{A} = c \cdot \overline{B}. \tag{4.9}$$

Then, because $\overline{\varnothing} = (0, \ldots, 0)$, it follows that $c \cdot \overline{\varnothing} = 0$. Because $\varnothing \prec X$, it follows from Equation 4.8 that $0 = c \cdot \overline{\varnothing} < c \cdot \overline{X}$.

Define the function $\mathbb{P}$ on $\mathcal{E}$ as follows: For each A in $\mathcal{E}$,

$$\mathbb{P}(A) = \frac{c \cdot \overline{A}}{c \cdot \overline{X}}.$$

Then $\mathbb{P}(X) = 1$, $\mathbb{P}(\varnothing) = 0$, and for all A and B in $\mathcal{E}$,

$$\text{if } A \prec B, \text{ then } c \cdot \overline{A} < c \cdot \overline{B}, \text{ and thus } \mathbb{P}(A) < \mathbb{P}(B),$$

and

$$\text{if } A \sim B, \text{ then } c \cdot \overline{A} = c \cdot \overline{B}, \text{ and thus } \mathbb{P}(A) = \mathbb{P}(B).$$

Suppose A and B are arbitrary elements of $\mathcal{E}$ such that $A \cap B = \varnothing$. Then

$$\overline{A \cup B} = \overline{A} + \overline{B},$$

and thus

$$\mathbb{P}(A \cup B) = \frac{c \cdot \overline{(A \cup B)}}{c \cdot \overline{X}} = \frac{c \cdot (\overline{A} + \overline{B})}{c \cdot \overline{X}} = \frac{c \cdot \overline{A} + c \cdot \overline{B}}{c \cdot \overline{X}} = \mathbb{P}(A) + \mathbb{P}(B).$$

The above shows that $\mathbb{P}$ is a probability representation of $\mathfrak{X}$. □

It is natural to inquire whether for some positive integer k the first k cancellation axioms imply the remaining cancellation axioms. The following theorem of Scott and Suppes (1958) provides the following answer:

Theorem 4.6 *Suppose $k \in \mathbb{I}^+$. Then there exists $\mathfrak{X} = \langle X, \mathcal{E}, \precsim \rangle$, where $\mathcal{E}$ is a boolean algebra of subsets of X and $\precsim$ is a reflexive relation on $\mathcal{E}$, such that $\mathfrak{X}$ satisfies the jth cancellation axiom for all j in $\mathbb{I}^+$, $j \leq k$, but $\mathfrak{X}$ does not satisfy the $(k+1)^{\text{th}}$ cancellation axiom.* □

4.3 $^\star\mathbb{R}$-probability Representations

This section considers probability representations for qualitative orderings on boolean algebras of sets. In order to accomplish this with the kind of generality achieved in the previous section for boolean algebras subsets of finite set, the notion of "representation" needs to be altered a little. There are two ways of doing this: (1) having probability representations be into the interval $^\star[0, 1]$ from some totally ordered field extension of the reals, or (2) having probability representations be real valued but weakening the condition,

$$\text{if } A \prec B \text{ then } \mathbb{P}(A) < \mathbb{P}(B),$$

to the condition

$$\text{if } A \prec B \text{ then } \mathbb{P}(A) \leq \mathbb{P}(B).$$

Both kinds of alterations provide better means for dealing with "sets of measure 0" of traditional probability theory. They were introduced in Narens (1974), and the theorems involving them of this and the following section are from Narens (1974).

Definition 4.7 $\mathbb{P}$ is said to be a $^\star\mathbb{R}$-*probability representation* of $\langle X, \mathcal{E}, \precsim\rangle$ if and only if

(1) $\mathcal{E}$ is a boolean algebra of subsets of X,

(2) $\precsim$ is a reflexive relation on $\mathcal{E}$,

(3) $^\star\mathfrak{R} = \langle {}^\star\mathbb{R}, \preceq, +, \cdot, 0, 1\rangle$ is a totally ordered field extension of the reals, and

(4) $\mathbb{P}$ is a function from $\mathcal{E}$ into the interval $^\star[0,1]$ of $^\star\mathfrak{R}$ such that the following four statements are true for all A and B in $\mathcal{E}$:

  (*i*) $\mathbb{P}(\varnothing) = 0$ and $\mathbb{P}(X) = 1$,
  (*ii*) If $A \prec B$, then $\mathbb{P}(A) < \mathbb{P}(B)$,
  (*iii*) If $A \sim B$, then $\mathbb{P}(A) = \mathbb{P}(B)$, and
  (*iv*) If $A \cap B = \varnothing$, then $\mathbb{P}(A \cup B) = \mathbb{P}(A) + \mathbb{P}(B)$. □

Definition 4.8 $\mathfrak{X} = \langle X, \mathcal{E}, \precsim\rangle$ is said to be a *qualitative probability structure* if and only if $\mathcal{E}$ is a boolean algebra of subsets of X, $\precsim$ is a reflexive relation on $\mathcal{E}$, and $\mathfrak{X}$ has a $^\star\mathbb{R}$-probability representation. □

Note that $\mathfrak{X} = \langle X, \mathcal{E}, \precsim\rangle$ is a qualitative probability structure if and only if it has either a $\mathbb{R}$-representation or a $^\star\mathbb{R}$-representation, because each $\mathbb{R}$-representation is a $^\star\mathbb{R}$-representation.

Definition 4.9 Let $\mathcal{D}$ be a boolean algebra of subsets of Y. Then $\mathcal{D}$ is said to be *finite* if and only if $\mathcal{D}$ is a finite set. *Note that according to this definition, Y is not required to be finite.* □

The following lemma is a well-known result about finite boolean algebras:

Lemma 4.2 *Suppose $\mathfrak{Y} = \langle Y, \mathcal{D}, \cup, \cap, - \rangle$ is a finite boolean algebra of subsets. Then there exist a finite set X and $\mathcal{E} = \wp(X)$ (the "power set of X," Convention 1.1) such that $\langle \mathcal{D}, \cup, \cap, - \rangle$ and $\langle \mathcal{E}, \cup, \cap, - \rangle$ are isomorphic.* $\square$

The proof of Lemma 4.2 is well-known and will only be outlined here with details left to the reader. An element A of $\mathcal{D}$ is called an "atom" if and only if $A \neq \varnothing$ and for all B in $\mathcal{D}$, if $\varnothing \subset B \subseteq A$, then $B = A$. Because $\mathcal{D}$ is finite, it is not difficult to show that for each element $C \neq \varnothing$ in $\mathcal{D}$, there exists an atom D in $\mathcal{D}$ such that $D \subseteq C$. Let X be the set of atoms in $\mathcal{D}$. For each A in $\mathcal{D}$ let

$$\widehat{A} = \text{the set of atoms } D \text{ in } \mathcal{D} \text{ such that } D \subseteq A.$$

Then X is finite because $\mathcal{D}$ is finite. Let $\mathcal{E} = \wp(X)$. It is not difficult to show that the function φ on $\mathcal{D}$ such that for each A in $\mathcal{D}, \varphi(A) = \widehat{A}$ is an isomorphism of $\langle \mathcal{D}, \cup, \cap, - \rangle$ onto $\langle \mathcal{E}, \cup, \cap, - \rangle$.

Convention 4.1 Suppose $\mathfrak{Y} = \langle Y, \mathcal{D}, \precsim \rangle$ is a finite boolean algebra of sets. Then $\mathfrak{Y}$ is said to *satisfy the finite cancellation axioms* if and only if there exist a finite set X and a binary relation $\precsim'$ on the power set of X, $\wp(X)$, such that (i) $\langle \mathcal{D}, \precsim, \cup, \cap, - \rangle$ and $\langle \wp(X), \precsim', \cup, \cap, - \rangle$ are isomorphic, and (ii) $\mathfrak{X} = \langle X, \wp(X), \precsim' \rangle$ satisfies the finite cancellation axioms as given in (Definition 4.6). Note by Lemma 4.2 such a structure $\mathfrak{X}$ exists.

Assuming the above notation of this convention, suppose $\mathfrak{Y}$ satisfies the finite cancellation axioms. Then $\mathfrak{X}$ satisfies the finite cancellation axioms, and by Theorem 4.5 let $\mathbb{P}$ be a probability representation of $\mathfrak{X}$. Let φ be an isomorphism from $\langle \mathcal{D}, \precsim, \cup, \cap, - \rangle$ onto $\langle \wp(X), \precsim', \cup, \cap, - \rangle$, and let $\mathbb{P}'$ be the function on $\mathcal{D}$ such that for each A in $\mathcal{D}$,

$$\mathbb{P}'(A) = \mathbb{P}(\varphi(A)).$$

Then, by isomorphism, $\mathbb{P}'$ is a probability representation of $\langle \mathcal{D}, \precsim, \cup, \cap, - \rangle$ and therefore of $\mathfrak{Y}$. $\square$

In summary, previous definitions, conventions, lemmas and theorems show that a finite boolean algebra of sets satisfies the finite cancellation axioms if and only if it has a probability representation.

Theorem 4.7 *Suppose the structure $\langle X, \mathcal{D}, \precsim \rangle$ has a ${}^{*}\mathbb{R}$-probability representation. Then for each finite boolean subalgebra $\mathcal{E}$ of $\mathcal{D}$, $\langle X, \mathcal{E}, \precsim \rangle$ satisfies the finite cancellation axioms.*

Proof. Same as Theorem 4.4 except that Lemma 4.2 is used to obtain a structure $\mathfrak{Z} = \langle Z, \wp(Z), \precsim' \rangle$ with a finite domain Z such that (i) $\langle \wp(Z), \cup, \cap, - \rangle$ is isomorphic to the finite boolean subalgebra $\langle \mathcal{E}, \cup, \cap, - \rangle$, and (ii) $\mathfrak{Z}$ is isomorphic to $\mathfrak{E} = \langle X, \mathcal{E}, \precsim \rangle$. Then "${}^*\mathbb{R}$-probability representation" for $\mathfrak{Z}$ is used in place of "probability representation" to show that $\mathfrak{Z}$ satisfies the finite cancellation axioms, which by isomorphism establishes that $\mathfrak{E}$ satisfies the finite cancellation axioms. □

Theorem 4.8 *Let $\mathfrak{X} = \langle X, \mathcal{E}, \precsim \rangle$, where $\mathcal{E}$ is a boolean algebra of subsets of X and $\precsim$ is a reflexive relation on X. Then the following two statements are equivalent:*

1. *$\mathfrak{X}$ is a qualitative probability structure.*
2. *$\varnothing \prec X$ and each finite boolean subalgebra of $\mathcal{E}$ satisfies the finite cancellation axioms (Convention 4.1).*

Proof. Assume Statement 1. Then by Theorem 4.7, each finite boolean subalgebra of sets of $\mathfrak{X}$ satisfies the finite cancellation axioms.

Assume Statement 2. Let

$$S = \{\alpha \mid \alpha \text{ is a finite boolean subalgebra of } \mathcal{E}\}$$

and

$$Y = \{\Delta \mid \Delta \text{ is a nonempty finite subset of } S\}.$$

For each α in S, let

$$\widehat{\alpha} = \{\Delta \mid \Delta \in Y \text{ and } \alpha \in \Delta\},$$

and

$$\mathcal{F} = \{\widehat{\alpha} \mid \alpha \in S\}.$$

Then $\mathcal{F}$ has the finite intersection property (Definition 3.8), because if $\widehat{\alpha_1}, \ldots, \widehat{\alpha_n}$ are in $\mathcal{F}$, then for $i = 1, \ldots, n$,

$$\{\alpha_1, \ldots, \alpha_n\} \in \widehat{\alpha_i},$$

and thus

$$\widehat{\alpha_1} \cap \cdots \cap \widehat{\alpha_n} \neq \varnothing.$$

Therefore, by Theorems 3.1 and 3.2, let $\mathcal{U}$ be an ultrafilter on Y such that $\mathcal{F} \subseteq \mathcal{U}$.

For each α in S, let $\precsim_\alpha$ be the restriction of $\precsim$ to α, and let $\mathfrak{X}_\alpha = \langle X, \alpha, \precsim_a \rangle$. By hypothesis, $\mathfrak{X}_\alpha$ satisfies the finite cancellation axioms for each α in S. Thus by the "summary" at the end of Convention 4.1, let $\mathbb{P}_\alpha$ be a probability representation for $\mathfrak{X}_\alpha$ for each α in S. For each A in $\mathcal{E}$, let F_A be the function from Y into $\mathbb{R}$ such that for each Δ in Y,

$$\beta = \text{the finite boolean subalgebra generated by } \bigcup \Delta$$

and

$$F_A(\Delta) = \begin{cases} \mathbb{P}_\beta(A) & \text{if } A \in \beta \\ = 0 & \text{if } A \notin \beta . \end{cases}$$

Let $\langle {}^\star\mathbb{R}, \leq, +, \cdot, 0, 1\rangle$ be the $\mathcal{U}$-ultrapower of $\langle \mathbb{R}, \leq, +, \cdot, 0, 1\rangle$. Let $\mathbb{P}$ be the function from $\mathcal{E}$ into ${}^\star\mathbb{R}$ such that for all A in $\mathcal{E}$, $\mathbb{P}(A) = F_A^\sim$.

Let A be an arbitrary element of $\mathcal{E}$. Let α in S be such that $A \in \alpha$. If $\Delta \in \widehat{\alpha}$, and $\beta =$ the finite boolean subalgebra generated by $\bigcup \Delta$, then

$$F_A(\Delta) = \mathbb{P}_\beta(A) \in [0, 1] .$$

Thus,

$$\{\Delta \mid F_A(\Delta) \in [0, 1]\} \supseteq \widehat{\alpha} \in \mathcal{U} ,$$

and therefore $F_A^\sim \in {}^\star[0, 1]$. Therefore, because A is an arbitrary element of $\mathcal{E}$, $\mathbb{P}$ is into ${}^\star[0, 1]$.

Let $\alpha \in S$, $\Delta \in \widehat{\alpha}$, and $\beta =$ the finite boolean subalgebra generated by $\bigcup \Delta$. Then,

$$F_\varnothing(\Delta) = \mathbb{P}_\beta(\varnothing) = 0 \text{ and } F_X(\Delta) = \mathbb{P}_\beta(X) = 1 ,$$

and thus,

$$\widehat{\alpha} \subseteq \{\Delta \mid F_\varnothing(\Delta) = 0\} \text{ and } \widehat{\alpha} \subseteq \{\Delta \mid F_X(\Delta) = 1\} .$$

Because $\widehat{\alpha} \in \mathcal{U}$, it follows that $F_\varnothing^\sim = 0$ and $F_X^\sim = 1$. Thus

$$\mathbb{P}(\varnothing) = F_\varnothing^\sim = 0 \text{ and } \mathbb{P}(X) = F_X^\sim = 1 .$$

Suppose A and B are arbitrary elements of $\mathcal{E}$ and $A \prec B$. Let α in S be such that A and B are in α. If $\Delta \in \widehat{\alpha}$, and $\beta =$ the finite boolean subalgebra generated by $\bigcup \Delta$, then,

$$F_A(\Delta) = \mathbb{P}_\beta(A) < \mathbb{P}_\beta(B) = F_B(\Delta) ,$$

and thus

$$\{\Delta \mid F_A(\Delta) < F_B(\Delta)\} \supseteq \widehat{\alpha} \in \mathcal{U} .$$

Therefore, $F_A^\sim < F_B^\sim$, and thus $\mathbb{P}(A) < \mathbb{P}(B)$.

By an argument similar to the above, it follows that for all C and D in $\mathcal{E}$, if $C \sim D$, the $\mathbb{P}(C) = \mathbb{P}(D)$.

Suppose A and B are arbitrary elements of $\mathcal{E}$ and $A \cap B = \varnothing$. Let α in S be such that A and B are in α. If $\Delta \in \widehat{\alpha}$, and $\beta =$ the finite boolean subalgebra generated by $\bigcup \Delta$, then,

$$F_{A \cup B}(\Delta) = \mathbb{P}_\beta(A \cup B) = \mathbb{P}_\beta(A) + \mathbb{P}_\beta(B) = F_A(\Delta) + F_B(\Delta),$$

and thus

$$\{\Delta \mid F_{A \cup B}(\Delta) = F_A(\Delta) + F_B(\Delta)\} \supseteq \widehat{\alpha} \in \mathcal{U},$$

and therefore, $\mathbb{P}(A \cup B) = \mathbb{P}(A) + \mathbb{P}(B)$. $\square$

Let $\mathfrak{X} = \langle X, \mathcal{E}, \precsim \rangle$ be a boolean algebra of subsets. One of the standard probabilistic interpretations of $\mathfrak{X}$ is that $\varnothing$ corresponds to impossibility and elements of X to states of the world, some possible and some impossible. The impossible states have no chance of occurring, while the possible states have some chance of occurring. It is often the case that the chances of some of the possible states occurring are vanishing small, and standard probability theory assigns probability 0 to these. In contrast, ${}^\star\mathbb{R}$-probability theory can assign positive infinitesimal values to such possible states. This better captures conceptually the idea of "possible state with a vanishing chance of occurring." A consequence of this is that ${}^\star\mathbb{R}$-probability theory provides a means for saying that "A has a vanishing more likely chance of occurring than B." Standard probability theory is only able to capture special cases of this, for example, when A and B are assigned the same probability and $A \supset B$. In ${}^\star\mathbb{R}$-probability theory, "A has a vanishing more likely chance of occurring than B," is simply and more completely captured by saying $\mathbb{P}(A) - \mathbb{P}(B)$ is infinitesimal.

In ${}^\star\mathbb{R}$-probability theory, conditional probabilities $A \mid B$ can be assigned to all A and B in $\mathcal{E}$ such that B is possible, that is, B contains a possible state of the world. This is not the case for standard probability theory when B is a set of measure 0. If one wants to "round off" ${}^\star\mathbb{R}$-probabilities to real numbers, that is, use ${}^\star\mathbb{R}$ to produce a real-valued probability function, then this is easily accomplished by taking standard parts (Definition 3.16). For conditional probabilities the same strategy works: Take ${}^\star\mathbb{R}$-conditional probabilities and then round off by takings standard parts. This method of rounding yields a coherent family of real valued conditional probabilities for all conditional events $A \mid B$ such that B is possible.

An important strategy for producing probability functions and probability representations is to use an axiomatization that implies the finite cancellation axioms to obtain a ${}^\star\mathbb{R}$-probability function, $\mathbb{P}$, which is then rounded off to obtain a probability function $\mathbb{Q}$. This strategy is used in the following theorem and section.

Theorem 4.9 *Let X be an infinite set and $\mathcal{E}$ be the power set of X. Then there exists a probability function $\mathbb{Q}$ on $\mathcal{E}$ such that for each x in X, $\mathbb{Q}(\{x\}) = 0$.*

Proof. Let $\precsim$ be the binary relation on $\mathcal{E}$ such that for all A and B in $\mathcal{E}$,

(1) $A \prec B$ if and only if (*i*) $A = \varnothing$ and $B = X$ or (*ii*) $A = \varnothing$ and $B = \{x\}$ for some x in X; and

(2) $A \sim B$ if and only if (*i*) $A = B$ or (*ii*) for some x and y in X, $A = \{x\}$ and $B = \{y\}$.

It is easy to verify that $\mathfrak{X} = \langle X, \mathcal{E}, \precsim \rangle$ satisfies the finite cancellation axioms. By Theorem 4.8, let $\mathbb{P}$ be a $^{\star}\mathbb{R}$-probability representation for $\mathfrak{X}$.

Let x and y be arbitrary elements of X. Then $\{x\} \sim \{y\}$. Therefore, $\mathbb{P}(\{x\}) = \mathbb{P}(\{y\})$. Because $\varnothing \prec \{x\}$ and $\mathbb{P}$ is a $^{\star}\mathbb{R}$-probability representation for $\mathfrak{X}$, $0 < \mathbb{P}(\{x\})$. It will be shown by contradiction that $\mathbb{P}(\{x\})$ is infinitesimal. Suppose not. Because $\mathbb{P}$ assigns to each singleton subset of X the same positive, non-infinitesimal value and there are infinitely many singleton subsets of X, it follows that for some finite union of singleton subsets of X, F, that $\mathbb{P}(F) > 1$, which contradicts $\mathbb{P}$ being a $^{\star}\mathbb{R}$-probability representation. Let $\mathbb{Q}$ be the function on $\mathcal{E}$ such that for all A in $\mathcal{E}$,

$$\mathbb{Q}(A) = {}^{\circ}\mathbb{P}(A)\,.$$

Then $\mathbb{Q}$ is a function from $\mathcal{E}$ into $[0, 1]$,

$$\mathbb{Q}(\varnothing) = 0 \text{ and } \mathbb{Q}(X) = 1\,,$$

and $\mathbb{Q}(\{x\}) = 0$ for each x in X (because $\mathbb{P}(\{x\})$ is infinitesimal). Let A and B be arbitrary elements of $\mathcal{E}$ such that $A \cap B = \varnothing$. Then

$$\mathbb{P}(A \cup B) = \mathbb{P}(A) + \mathbb{P}(B)\,,$$

and thus by Theorem 3.10,

$$\mathbb{Q}(A \cup B) = \mathbb{Q}(A) + \mathbb{Q}(B)\,. \quad \square$$

4.4 Weak Probability Representations

When applied to ${}^\star\mathbb{R}$-representations of a qualitative probability structure with a qualitative ordering of event $\precsim$, rounding off by taking standard parts yields representations φ that sometimes require $A \prec B$ to be represented as $\varphi(A) = \varphi(B)$. Rounding off in this case yields the following concept of "representation:"

Definition 4.10 $\mathbb{P}$ is said to be a *weak probability representation* of the structure $\langle X, \mathcal{E}, \precsim \rangle$ if and only if

(1) $\mathcal{E}$ is a boolean algebra of subsets of X,

(2) $\precsim$ is a reflexive relation on $\mathcal{E}$, and

(3) $\mathbb{P}$ is a function from $\mathcal{E}$ into the closed interval $[0,1]$ of $\langle \mathbb{R}, \leq \rangle$ such that the following four statements are true for all A and B in $\mathcal{E}$:

   (*i*) $\mathbb{P}(\varnothing) = 0$ and $\mathbb{P}(X) = 1$,

   (*ii*) If $A \prec B$, then $\mathbb{P}(A) \leq \mathbb{P}(B)$,

   (*iii*) If $A \sim B$, then $\mathbb{P}(A) = \mathbb{P}(B)$, and

   (*iv*) If $A \cap B = \varnothing$, then $\mathbb{P}(A \cup B) = \mathbb{P}(A) + \mathbb{P}(B)$. □

The above definition of "weak probability representation" is formulated entirely in terms standard probabilistic concepts. However, one important way of producing weak probability representations is through ${}^\star\mathbb{R}$-representations:

Theorem 4.10 *Suppose* $\mathfrak{X} = \langle X, \mathcal{E}, \precsim \rangle$ *is a qualitative probability structure. Then* $\mathfrak{X}$ *has a weak probability representation.*

Proof. By Definition 4.8, let $\mathbb{P}$ be a ${}^\star\mathbb{R}$-representation of $\mathfrak{X}$. Because $\mathbb{P}$ is into ${}^\star[0,1]$, $\mathbb{P}(A)$ is finite for each $A \in \mathcal{E}$. Let $\mathbb{Q}$ be the function on $\mathcal{E}$ such that for all A in $\mathcal{E}$,

$$\mathbb{Q}(A) = {}^\circ\mathbb{P}(A) .$$

Then $\mathbb{Q}$ is a function from $\mathcal{E}$ into $[0,1]$ and

$$\mathbb{Q}(\varnothing) = 0 \text{ and } \mathbb{Q}(X) = 1 .$$

Let A and B be arbitrary elements of $\mathcal{E}$. If $A \prec B$, then $\mathbb{P}(A) < \mathbb{P}(B)$, and thus by Theorem 3.9, $\mathbb{Q}(A) \leq \mathbb{Q}(B)$. If $A \sim B$, then $\mathbb{Q}(A) = \mathbb{Q}(B)$. If $A \cap B = \varnothing$, then

$$\mathbb{P}(A \cup B) = \mathbb{P}(A) + \mathbb{P}(B) ,$$

and thus by Theorem 3.10,

$$\mathbb{Q}(A \cup B) = \mathbb{Q}(A) + \mathbb{Q}(B)\,.$$

Therefore, $\mathbb{Q}$ is a weak probability representation of $\mathfrak{X}$. $\square$

The following theorem characterizes a situation in which the unique weak probability representation of a qualitative probability structure is approximated by the weak probability representations of a sufficiently large substructure.

Theorem 4.11 *Suppose $\mathfrak{X} = \langle X, \mathcal{E}, \precsim\rangle$ is a qualitative probability structure with a unique probability representation $\mathbb{P}$, $\mathfrak{X}_i = \langle X, \mathcal{E}_i, \precsim_i\rangle$ is a sequence of qualitative probability structures with (not necessarily unique) weak probability representations $\mathbb{P}_i$, and for each i in $\mathbb{I}^+$ suppose*

$$\mathcal{E}_i \subseteq \mathcal{E}_{i+1} \ \ \textit{and} \ \ \precsim_i \subseteq \precsim_{i+1}$$

and

$$\mathcal{E} = \bigcup_{i=1}^{\infty} \mathcal{E}_i \ \ \textit{and} \ \ \precsim = \bigcup_{i=1}^{\infty} \precsim_i \ .$$

Then $\mathbb{P} = \lim_{i\to\infty} \mathbb{P}_i$.

Proof. Suppose $\mathbb{P} \neq \lim_{i\to\infty} \mathbb{P}_i$. A contradiction will be shown. Because $\mathbb{P} \neq \lim_{i\to\infty} \mathbb{P}_i$, let $G \in \mathcal{E}$, $r \in \mathbb{R}^+$, and J an infinite subset of $\mathbb{I}^+$ be such that for each j in J,

$$r < |\mathbb{P}(G) - \mathbb{P}_j(G)|\,.$$

Let $\mathcal{F}$ be the filter of co-finite subsets of J (Definition 3.5). By Theorem 3.2 let $\mathcal{U}$ be an ultrafilter on J such that $\mathcal{F} \subseteq \mathcal{U}$. Let $\langle {}^\star\mathbb{R}, \leq, +, \cdot\rangle$ be the $\mathcal{U}$-ultrapower of $\langle \mathbb{R}, \leq, +, \cdot\rangle$. For each A in $\mathcal{E}$, let $\mathbb{Q}_A$ be the function from J into $\mathbb{R}$ such that for each j in J,

$$\mathbb{Q}_A(j) = \begin{cases} \mathbb{P}_j(A) & A \in \mathcal{E}_j \\ 0 & A \notin \mathcal{E}_j\,. \end{cases}$$

Then for each A in $\mathcal{E}$, $Q_A^{\sim}$ is in ${}^\star[0,1]$. Let $\mathbb{Q}$ be the function on $\mathcal{E}$ such that for each A in $\mathcal{E}$,

$$\mathbb{Q}(A) = {}^\circ\mathbb{Q}_A^{\sim}\,.$$

Then for each A in $\mathcal{E}$, $\mathbb{Q}(A)$ is in $[0,1]$. It will now be shown that $\mathbb{Q}$ is a weak probability representation of $\mathfrak{X}$.

Because

$$\{j \in J \mid \mathbb{Q}_X(j) = 1\} = \{j \in J \mid P_j(X) = 1\} = J \in \mathcal{U},$$

it follows that

$$\mathbb{Q}(X) = {}^\circ\mathbb{Q}_X^\sim = 1.$$

Similarly,

$$\mathbb{Q}(\varnothing) = 0.$$

Suppose A and B are arbitrary elements of $\mathcal{E}$ and $A \prec B$. Let n in J be such that for each j in J, if $n \leq j$ then A and B are in $\mathcal{E}_j$. Then

$$\begin{aligned}\{j \mid j \in J \text{ and } \mathbb{Q}_A(j) < \mathbb{Q}_B(i)\} &\supseteq \{j \mid j \in J\,,\ n \leq j, \text{ and } \mathbb{P}_j(A) < \mathbb{P}_j(B)\} \\ &= \{j \mid j \in J \text{ and } n \leq j\} \in \mathcal{U}.\end{aligned}$$

Therefore, as elements of ${}^\star\mathbb{R}$, $\mathbb{Q}_A^\sim < \mathbb{Q}_B^\sim$, and thus

$$\mathbb{Q}(A) = {}^\circ\mathbb{Q}_A^\sim \leq {}^\circ\mathbb{Q}_B^\sim = \mathbb{Q}(B).$$

Similarly, if C and D are arbitrary elements of $\mathcal{E}$ such that $C \sim D$, then $\mathbb{Q}(C) = \mathbb{Q}(D)$.

Suppose E and F are in $\mathcal{E}$ and $E \cap F = \varnothing$. Let m in J be such that for each j in J, if $m \leq j$ then E and F are in $\mathcal{E}_j$. Then

$$\begin{aligned}&\{j \mid j \in J \text{ and } \mathbb{Q}_{E\cup F}(j) = \mathbb{Q}_E(j) + \mathbb{Q}_j(F)\} \\ &\quad \supseteq \{j \mid j \in J,\ m \leq j, \text{ and } \mathbb{P}_j(E \cup F) = \mathbb{P}_j(E) + \mathbb{P}_j(F)\} \\ &\quad = \{j \mid j \in J \text{ and } m \leq j\} \in \mathcal{U}.\end{aligned}$$

Thus,

$$\mathbb{Q}_{E\cup F}^\sim = \mathbb{Q}_E^\sim + \mathbb{Q}_F^\sim$$

and therefore,

$$\mathbb{Q}(E \cup F) = {}^\circ(\mathbb{Q}_E^\sim + \mathbb{Q}_F^\sim) = {}^\circ\,\mathbb{Q}_E^\sim + {}^\circ\,\mathbb{Q}_F^\sim = \mathbb{Q}(E) + \mathbb{Q}(F).$$

The above shows that $\mathbb{Q}$ is a weak probability representation of $\mathfrak{X}$. Recall that at the beginning of this proof that G and r were chosen so that for each j in J, $r \leq |P_j(G) - \mathbb{P}(G)|$. From this it follows that

$$\begin{aligned}&\{j \mid j \in J \text{ and } r \leq |\mathbb{Q}_G(j) - \mathbb{P}(G)|\} \\ &\quad = \{j \mid j \in J \text{ and } r \leq |\mathbb{P}_j(G) - \mathbb{P}(G)|\} = J \in \mathcal{U}.\end{aligned}$$

Thus in $\langle {}^\star\mathbb{R}, \leq, +, \cdot\rangle$,

$$r \leq |\mathbb{Q}_G^\sim - \mathbb{P}(G)|.$$

Because $\mathbb{Q}(G) = {}^{\circ}\mathbb{Q}^{\sim}_{G}$, it follows that

$$\frac{r}{2} \leq |\mathbb{Q}(G) - \mathbb{P}(G)|\,.$$

Thus $\mathbb{Q} \neq \mathbb{P}$, contradicting that $\mathbb{P}$ is a unique weak probability representation of $\mathfrak{X}$. □

4.5 Independence

This section is based on results of Luce and Narens 1978 article, "Qualitative Independence in Probability Theory." They begin their article by noting,

> Probability theory is measure theory specialized by assumptions having to do with stochastic independence. Delete from probability and statistics those theorems that explicitly or implicitly (e.g., by postulating a random sample) invoke independence, relatively little remains. Or attempt to estimate probabilities from data without assuming that at least certain observations are independent, and little results. Everyone who has worked with or applied probability is keenly aware of the importance of stochastic independence; experimenters go to some effort to ensure, and to check, that repeated observations are independent. Kolmogorov (1933, 1950) wrote:
>
> > The concept of mutual *independence* of two or more experiments holds, in a certain sense, a central position in the theory of probability. (p. 8).
> >
> > In consequence, one of the most important problems in the philosophy of the natural sciences is — in addition to the well-known one regarding the essence of the concept of probability itself — to make precise the premises which would make it possible to regard any give real events as independent. (p. 9).
>
> ... Since it is easy to give examples of [qualitative probability structures with nonunique probability representations], it is clear that stochastic independence cannot be defined in terms of [the qualitative ordering $\precsim$ on events]. ... In this connection, two quotations from Fine (1973) are relevant:
>
> > ... we must be cognizant of the fact that invocations of [stochastic independence] are usually not founded

> upon empirical or objective knowledge of probabilities. Quite the contrary. Independence is adduced to permit us to simplify and reduce the family of possible probabilistic descriptions for a given experiment (p. 80).
>
> Given our views as to the problems encountered in assessing probability, we do not favor a purely probability-based definition [of independence] (p. 141).
>
> *(Luce and Narens, 1978, pp. 225–226)*

Consistent with the above quotations, the concept of qualitative independence developed in this section is that one can decide whether two events A and B are independent, in symbols, $A \perp B$, without reference to their numerical probabilities. This idea leads to the following definition.

Definition 4.11 $\langle \mathcal{X}, \precsim, \perp, X \rangle$ is said to be a *qualitative probability structure with independence* if and only if $\langle X, \mathcal{X}, \precsim \rangle$ is a qualitative probability structure, $\perp$ is a binary relation on $\mathcal{X}$, and for each probability representation $\mathbb{P}$ of $\langle X, \mathcal{X}, \precsim \rangle$ and each A and B in $\mathcal{X}$,

$$A \perp B \text{ implies } \mathbb{P}(A \cap B) = \mathbb{P}(A) \cdot \mathbb{P}(B)\,. \quad \square$$

4.5.1 Independent Experiments

Luce and Narens (1978) notes that,

> In the case of independent experiments, one usually finds the discussion [regarding independence] cast in terms of two or more independent random variables, as in a random sample. The usual probabilistic approach assumes that the experiments underlying the two random variables are each run, that a joint probability distribution exists over the various pairs of events, and that the relevant pairs induced by the random variables are stochastically independent as events. In practice, however, what one does is attempt to devise experimental realizations for which there are ample structural reasons for believing the two experiments are independent of each other in the sense that knowledge of one does not affect the other. This concept of independence is discussed at length by Keynes (1962), see especially Ch. XVI. *(p. 226)*

The following definition is intended to capture the idea of the same experiment, $\mathcal{E}$, being repeated twice, $\mathcal{E} \times \mathcal{E}$. (To be strictly correct, the two occurrences of $\mathcal{E}$ should be treated as isomorphic copies and the notation should reflect this, but for reasons of exposition, this subtlety is ignored, leading to a harmless abuse of notation.) In the following definition, the qualitative ordering of likelihood of occurrence, $\precsim$, is intended to capture qualitatively some of the key properties of a joint probability distribution on $\mathcal{E} \times \mathcal{E}$.

Definition 4.12 Suppose $\mathcal{E}$ is a boolean algebra of subsets of X and $\precsim$ is a binary relation on $\mathcal{E} \times \mathcal{E}$. Then $\langle \mathcal{E} \times \mathcal{E} \ \precsim, X\rangle$ is said to be an *independent joint qualitative probability structure*; if and only if the following nine axioms hold for all nonempty A, B, C, D, E, and F in $\mathcal{E}$:

(1) *(Weak Ordering)* $\precsim$ is a weak ordering and $(\varnothing, \varnothing) \sim (\varnothing, A) \prec (B, C)$.

(2) *(Independence)* $(A, C) \precsim (B, C)$ if and only if $(A, D) \precsim (B, D)$.

(3) *(Symmetry)* $(A, B) \sim (B, A)$.

(4) *(Distributivity)* If $A \cap B = C \cap D = \varnothing$, $(A, E) \precsim (C, F)$, and $(B, E) \precsim (D, F)$, then

$$(A \cup B, E) \precsim (C \cup D, F) . \tag{4.10}$$

Moreover, the conclusion, Equation 4.10, is $\prec$ if either hypothesis is $\prec$.

Define the binary relation $\precsim_{\mathcal{E}}$ on $\mathcal{E}$ as follows: For all G and H in $\mathcal{E}$, $G \precsim_{\mathcal{E}} H$ if and only if for all $K \neq \varnothing$ in $\mathcal{E}$, $(G, K) \precsim (H, K)$. Then it follows from Axioms (1), (2), and (3) that $\precsim_{\mathcal{E}}$ is a weak ordering on $\mathcal{E}$.

(5) *(Sure Event and Null Event)* $\varnothing \prec_{\mathcal{E}} X$ and $\varnothing \precsim A \precsim X$.

(6) (*Archimedean*) Every sequence A_i of the following form is finite:

$$\varnothing \prec_{\mathcal{E}} C \prec_{\mathcal{E}} B,\ \varnothing \prec_{\mathcal{E}} A_1, \ \text{and} \ (A_{i+1}, C) \sim_{\mathcal{E}} (A_i, B) .$$

(7) There exists U in $\mathcal{E}$ such that $(U, X) \sim (A, B)$.

(8) If $A \cap B = \varnothing$ and $(C, D) \precsim (A, B)$, then there exists C' and D' in $\mathcal{E}$ such that

$$C \precsim_{\mathcal{E}} C',\ D \precsim_{\mathcal{E}} D', \ \text{and} \ C' \cap D' = \varnothing .$$

(9) If $B \precsim_{\mathcal{E}} A$, then there exists A' in $\mathcal{E}$ such that $A' \sim_{\mathcal{E}} A$ and $B \subseteq A'$.

□

Theorem 4.12 *Suppose $\mathcal{E}$ is a boolean algebra of subsets and $\langle \mathcal{E}\times\mathcal{E}, \precsim, X\rangle$ is an independent joint qualitative probability structure. Then there exists a unique function $\mathbb{P}$ on $\mathcal{E}$ such that the following two statements hold:*

1. *$\mathbb{P}$ is a finitely additive probability function on $\langle \mathcal{E}, \cup, \cap, {}^{-}, X, \varnothing\rangle$.*

2. *For all A, B, C, and D in $\mathcal{E}$,*

$$(A, B) \precsim (C, D) \text{ iff } \mathbb{P}(A) \cdot \mathbb{P}(B) \leq \mathbb{P}(C) \cdot \mathbb{P}(D) .$$

Proof. Luce and Narens (1978). □

Note that axioms (1) to (6) in the definition of an independent joint qualitative probability structure (Definition 4.12) are necessary for the existence of the probability function in Theorem 4.12.

Although it is natural to view elements of $\mathcal{E} \times \mathcal{E}$ as "events," the event space for which they are events has not been characterized. The following definition provides such a characterization.

Definition 4.13 Suppose $\mathbb{P}$ is a finitely additive probability function on the boolean algebra $\langle \mathcal{E}, \cup, \cap, {}^{-}, X, \varnothing\rangle$.

By definition, let $\widehat{\mathcal{E}\times\mathcal{E}}$ be the set of all subsets of $X \times X$ which are unions of finitely many disjoint subsets of $\mathcal{E} \times \mathcal{E}$.

Define $\widehat{\mathbb{P}}$ on $\widehat{\mathcal{E}\times\mathcal{E}}$ as follows: For each E in $\widehat{\mathcal{E}\times\mathcal{E}}$ with

$$E = \bigcup_{i=1}^{n} (A_i, B_i) ,$$

where for $i, j = 1, \ldots, n$, $i \neq j$, (A_i, B_i) and (A_j, B_j) in $\mathcal{E}\times\mathcal{E}$, and $(A_i, B_i) \cap (A_j, B_j) = \varnothing$,

$$\widehat{\mathbb{P}}(E) = \sum_{i=1}^{n} \mathbb{P}(A_i) \cdot \mathbb{P}(B_i) . \quad \square$$

Theorem 4.13 *Suppose $\mathbb{P}$ is a finitely additive probability function on the boolean algebra $\mathfrak{E} = \langle \mathcal{E}, \cup, \cap, {}^{-}, X, \varnothing\rangle$ and $\widehat{\mathcal{E}\times\mathcal{E}}$ and $\widehat{\mathbb{P}}$ are as in Definition 4.13. Then the following five statements are true:*

1. *$\widehat{\mathcal{E}\times\mathcal{E}}$ is a boolean algebra of subsets of $X \times X$.*

2. *$\widehat{\mathcal{E}\times\mathcal{E}}$ is the minimal boolean algebra of subsets of $X \times X$ that contains $\widehat{\mathcal{E}\times\mathcal{E}}$.*

3. *$\widehat{\mathbb{P}}$ is a finitely additive probability function on $\mathfrak{E}$.*

4. *$\mathfrak{E}$ is the unique boolean algebra with the property that for all (A, B) in $\mathcal{E} \times \mathcal{E}$, $\widehat{\mathbb{P}}[(A, B)] = \mathbb{P}(A) \cdot \mathbb{P}(B)$.*

5. $\widehat{\mathbb{P}}[(A, X) \cap (X, B)] = \widehat{\mathbb{P}}[(A, X)] \cdot \widehat{\mathbb{P}}[(X, B)]$.

Proof. The proof follows from Definition 4.13 in a straightforward manner. The details are given in Luce and Narens (1978). □

Luce and Narens (1978) makes the following comment about the above two theorems.

> The literature of numerical representations of qualitative probability structures is not really satisfactory for the following reason. In order for the representation to be unique (with $\mathbb{P}(X) = 1$), one is forced to postulate very strong solvability conditions. Without such conditions, such as in the finite case, not only do several representations exist but their relations to one another are difficult to characterize. But the real issue centers around nonuniqueness which is clearly incompatible with independence being formulated as $\mathbb{P}(A \cap B) = \mathbb{P}(A) \cdot \mathbb{P}(B)$. This has led many of us to believe that were we to introduce an appropriate qualitative notion of Independence, it could be used to force uniqueness of the representation. So far, however, this has not proved successful. *(p. 231)*

In addition to the case of independent experiments Luce and Narens (1978) also provide qualitative axiomatizations for the case of independent events (using Axioms 4.1 to 4.3 below). The following provides a sketch of the main idea of their axiomatizations and proofs of the corresponding representation and uniqueness theorems (Theorems 4.14 and 4.15 below). This idea reveals what I consider to be the key qualitative property of independence and provides a rationale as to why one should expect the probabilities of independent events to multiply.

Consider first a situation of casino gambling, say roulette. Let $\mathcal{E}$ be the boolean algebra generated by the various combinations events that can occur on a given turn of the wheel (e.g., the event, 2 or 4 or an odd number coming up). Let $\precsim_{\mathcal{E}}$ be a rational agent's likelihood ordering on the events in $\mathcal{E}$. Let L be an event that the agent believes to be unrelated to roulette, say the basketball team the Los Angeles Lakers winning their next game. Let $\precsim_{L \cap \mathcal{E}}$ be the agent's likelihood ordering over the boolean algebra of events $L \cap \mathcal{E} = \{L \cap E \,|\, E \in \mathcal{E}\}$. Then if the agent considers L to be "independent" of the events in $\mathcal{E}$, that is, $L \perp E$ for each E in $\mathcal{E}$, then "rationality" should require that for all E and F in $\mathcal{E}$,

$$E \precsim_{\mathcal{E}} F \text{ iff } L \cap E \precsim_{L\cap\mathcal{E}} L \cap F, \qquad (4.11)$$

or in other words, the function φ defined on $\mathcal{E}$ by $\varphi(A) = L \cap A$ is an isomorphism of $\langle \mathcal{E}, \precsim_{\mathcal{E}} \rangle$ onto $\langle \mathcal{E}_{L\cap\mathcal{E}}, \precsim_{\mathcal{E}} \rangle$.

Instead of roulette events, now consider the case where

- $\langle \mathcal{E}, \precsim_{\mathcal{E}}, X \rangle$ is an arbitrary qualitative probability structure with a unique finitely additive probability representation $\mathbb{P}_{\mathcal{E}}$,
- L is an arbitrary event such that $L \perp \mathcal{E}$ and $L \notin \mathcal{E}$,
- $\precsim_{\mathcal{E}}$ and $\precsim_{L\cap\mathcal{E}}$ are the agent's likelihood orderings on respectively $\mathcal{E}$ and $L \cap \mathcal{E}$ $(= \{L \cap E \,|\, E \in \mathcal{E})$,
- and Equation 4.11 is satisfied.

Let $\mathcal{F}$ be the smallest boolean algebra containing $\mathcal{E}$ and $L \cap \mathcal{E}$ and let $\mathbb{P}$ be the agent's finitely additive probability function on $\mathcal{F}$. Because the agent is rational, we assume that $\mathbb{P} = \mathbb{P}_{\mathcal{E}}$ on $\mathcal{E}$. Then, using Equation 4.11 and the facts that for all E and F in $\mathcal{E}$,

$$L \cap (E \cup F) = (L \cap E) \cup (L \cap F) \text{ and } L \cap (E \cap F) = (L \cap E) \cap (L \cap F),$$

it easily follows that the function on $\mathbb{P}'$ on $\mathcal{E}$ such that for all G in $\mathcal{E}$

$$\mathbb{P}'(G) = \frac{1}{\mathbb{P}(L \cap X)} \cdot \mathbb{P}(L \cap G), \qquad (4.12)$$

is a finitely additive probability representation on $\mathcal{E}$. Because by hypothesis $\mathbb{P}_{\mathcal{E}}$ is the unique finitely additive probability representation on $\mathcal{E}$, it follows that $\mathbb{P}_{\mathcal{E}} = \mathbb{P}'$. Because $\mathbb{P} = \mathbb{P}_{\mathcal{E}}$ on $\mathcal{E}$, it then follows from $\mathbb{P}(L \cap X) = \mathbb{P}(L)$ and Equation 4.12 that for each G in $\mathcal{E}$,

$$\mathbb{P}(L) \cdot \mathbb{P}(G) = \mathbb{P}(L \cap X) \cdot \mathbb{P}(G) = \mathbb{P}(L \cap G),$$

showing that for each G in $\mathcal{E}$, L and G are independent in the Kolmogorov sense.

In my opinion the above discussion reveals Equation 4.11 to be the essence of probabilistic concept of independence.

Luce and Narens (1978) use the ideas in the above discussion to provide a qualitative axiomatization for the Kolmogorov concept of independence. Specifically, their axiomatization consists of the following:

They start with a qualitative probability structure $\langle \mathcal{E}, \precsim, X \rangle$ with a unique finitely additive probability representation $\mathbb{P}$. (They comment: "See Krantz et al., 1971, Ch. 5 for various sufficient conditions for $\mathbb{P}$ to exists.") They add another qualitative relation $\perp$ "to be interpreted as independence of events." For A in $\mathcal{E}$ and boolean subalgebras $\mathcal{H}$ of $\mathcal{E}$ they define $A \perp \mathcal{H}$ in a manner like above. "To insure that $\perp$ is a sufficiently rich relation" they impose the following axiom:

Axiom 4.1 *There exists a set $\mathcal{E}'$ of subsets of X with the following four properties:*

(i) *$\mathcal{E}'$ is a boolean algebra of sets.*

(ii) *$\mathcal{E}' \subseteq \mathcal{E}$.*

(iii) *For all A in $\mathcal{E}$ there exists A' in $\mathcal{E}'$ such that $A \sim A'$.*

(iv) *For all A' in $\mathcal{E}'$ there exists A in $\mathcal{E}$ such that $A \sim A'$ and $A \perp \mathcal{E}'$.* □

They then characterize the "interlock between $\perp$, $\precsim$, and $\cap$" as the following axiom:

Axiom 4.2 *For all A, B, C, and D in $\mathcal{E}$, suppose $A \perp B$, $C \perp D$, and $A \sim C$. Then*

$$B \precsim D \text{ iff } A \cap B \precsim C \cap D . \quad \square$$

They then show the following theorem.

Theorem 4.14 *Assume Axioms 4.1 and 4.2. Then for all A and B in $\mathcal{E}$,*

$$A \perp B \text{ implies } \mathbb{P}(A \cap B) = \mathbb{P}(A) \cdot \mathbb{P}(B) . \quad \square$$

Luce and Narens show that Theorem 4.14 can be strengthen to capture the Kolmogorov concept of independence by including the following axiom:

Axiom 4.3 *For all A, B, C, and D in $\mathcal{E}$, if $A \perp B$, $A \sim C$, $B \sim D$, and $A \cap B \sim C \cap D$, then $C \perp D$.* □

Theorem 4.15 *Assume Axioms 4.1, 4.2, and 4.3. Then for all A and B in $\mathcal{E}$,*

$$A \perp B \text{ iff } \mathbb{P}(A \cap B) = \mathbb{P}(A) \cdot \mathbb{P}(B) . \quad \square$$

Chapter 5

Qualitative Utility

5.1 Brief Overview

This chapter explores a very small portion of the extensive literature of inferring subjective probabilities through gambling behavior. The widely used SEU model of utility theory is described in Section 5.2. A qualitative axiomatization of it is presented for the highly restrictive, but still interesting, case of two outcome gambles where one outcome is receiving nothing. An empirical study is presented showing the failure of SEU for this case.

Section 5.3 discusses two generalizations of SEU—Dual Bilinear Utility Theory and Rank Dependent Utility Theory. The generalizations agree with each other for gambles involving two pure outcomes, but differ when more outcomes are involved. They incorporate features that have psychological or economic importance that are excluded by SEU. The relevance of utility models for inferring subjective probabilities is also discussed. It is argued that the use of such models for determining subjective probabilities is unfounded unless additional theory or empirical results are provided.

5.2 Subjective Expected Utility

Convention 5.1 Throughout this chapter the following notation and conventions are observed:

- X stands for the sure event.
- (a, X) stands for the *sure gamble* of receiving the object of value a for certain.
- Non-sure gambles are called *proper gambles.* Proper gambles have the form $G = (a_1, A_1; \ldots; a_n, A_n)$, where $a_1, \ldots, a_n$ are objects of value

and $A_1, \ldots, A_n$ is a partition X. The objects of value, $a_1, \ldots, a_n$, of G need not be distinct. However, by the definition of "partition" (Definition 1.6), the events, $A_1, \ldots, A_n$ are distinct and non-null. $G = (a_1, A_1; \ldots; a_n, A_n)$ stands for receiving a_1 if A_1 occurs, ..., a_n if A_n occurs.

- It is allowed that some of the objects of value of a proper gamble G can be proper gambles. In such situations G is called a *compound gamble.* A gamble that is neither a compound gamble nor a sure gamble is called a *simple gamble.* The objects of value in a simple gamble are called *pure outcomes.*
- V denotes the set of pure outcomes. □

Definition 5.1 $\mathfrak{G} = \langle \mathcal{G}, \precsim \rangle$ is said to be a *gambling utility structure* if and only if $\mathcal{G}$ is a nonempty set of gambles and $\precsim$ is a binary relation on $\mathcal{G}$.

Let $\mathfrak{G} = \langle \mathcal{G}, \precsim \rangle$ be a gambling utility structure. $\precsim$ is called the *preference relation* of $\mathfrak{G}$. In the intended interpretation, "$G \precsim H$" is interpreted as the "the decision maker evaluates that G has less than or as much value (to him or her) as H." The *set of objects of value of* $\mathfrak{G}$ is

$$\{a \mid a \text{ is an object of value for some } G \in \mathcal{G}\}.$$

The *set of events of* $\mathfrak{G}$ is

$$\{A \mid A \text{ is an event for some } G \in \mathcal{G}\}. \quad \square$$

Definition 5.2 Let $\mathfrak{G} = \langle \mathcal{G}, \precsim \rangle$ be a gambling utility structure. Then $\langle u, \mathbb{P} \rangle$ is said to be a *SEU (Subjective Expected Utility) representation* for $\mathfrak{G}$ if and only if

(*i*) u is a function from the set of objects of value of $\mathfrak{G}$ into $\mathbb{R}$,

(*ii*) $\mathbb{P}$ is a function from the set of events of $\mathfrak{G}$ into $[0, 1]$ such that

(1) $\mathbb{P}(X) = 1$ and $\mathbb{P}(\varnothing) = 0$, and

(2) for all events A and B of $\mathfrak{G}$ such that $A \cap B = \varnothing$,

$$\mathbb{P}(A \cup B) = \mathbb{P}(A) + \mathbb{P}(B),$$

and

(*iii*) for all $G = (a_1, A_1; \ldots; a_n, A_n)$ and $H = (b_1, B_1; \ldots; b_m, B_m)$ in $\mathcal{G}$,

$$G \precsim H \text{ iff } \sum_{i=1}^{n} u(a_i)\mathbb{P}(A_i) \le \sum_{j=1}^{m} u(b_j)\mathbb{P}(B_j). \quad \square \tag{5.1}$$

Definition 5.3 Let $\mathfrak{G} = \langle \mathcal{G}, \precsim \rangle$ be a gambling utility structure and $\langle u, \mathbb{P} \rangle$ be a SEU representation for $\mathfrak{G}$. Then u is said to be a *$\precsim$-induced utility function* (on the set of objects of value of $\mathfrak{G}$), and $\mathbb{P}$ is said to be a *$\precsim$-induced probability function* (on the set of events of $\mathfrak{G}$). □

Note that in Definition 5.3, if u is a $\precsim$-induced utility function, then $ru + s$ is also a $\precsim$-induced utility function for each $r \in \mathbb{R}^+$ and each $s \in \mathbb{R}$.

In the literature various qualitative axioms are presented about a gambling structure $\mathfrak{G}$ that yield a *representation and uniqueness theorem* of the following form:

(1) (Representation) *There exists a SEU representation for $\mathfrak{G}$; and*

(2) (Uniqueness) *for all SEU representations $\langle u, \mathbb{P} \rangle$ and $\langle v, \mathbb{Q} \rangle$ for $\mathfrak{G}$, $\mathbb{P} = \mathbb{Q}$ and there exist $r \in \mathbb{R}^+$ and $s \in \mathbb{R}$ such that $u = rv + s$.*

The only theorem of this form presented in the chapter involves the special case of simple gambles of the form $(a, A; 0, X - A)$, that is, gambles of receiving a of A occurs and nothing if A does not occur.

Observe that a representation and uniqueness theorem for SEU require that the $\precsim$-induced probability function $\mathbb{P}$ is unique and the $\precsim$-induced utility functions form an interval scale $\mathcal{U}$ (Definition 1.12). Most researchers in utility theory identify $\mathcal{U}$ as a scale of functions that measures the Decision Maker's subjective value or *utility* for objects of value, and identify $\mathbb{P}$ as a function that measures the Decision Maker's subjective probability for events. I believe that such interpretations are misleading:

SEU directly measures (or "represents") the Decision Maker's subjective ordering of gambles, $\precsim$. The $\precsim$-induced probability function, $\mathbb{P}$, and the interval scale, $\mathcal{U}$, of $\precsim$-induced value functions are only derived from $\precsim$; they do not directly measure uncertainty or value. It is reasonable to suppose that the Decision Maker has a subjective belief function on events, $\mathbb{B}$, that *directly* measures his or her degrees of belief for events. Theoretically, $\mathbb{B}$ could result from a qualitative ordering $\precsim_{\mathbb{B}}$ on the events of $\mathfrak{G} = \langle \mathcal{G}, \precsim \rangle$ that is elicited from the Decision Maker without reference to the gambles in $\mathcal{G}$ or any other gambles. If $\mathbb{B}$ is a probability function, then it is reasonable to call $\mathbb{B}$ a *subjective probability function*. Similarly, theoretically an interval scale family of utility functions $\mathcal{V}$ that directly measures the value (to the Decision Maker) of elements of V can be directly elicited without reference to gambles.[1] $\mathcal{V}$ is called the Decision Maker's *family of subjective utility*

[1]One method for accomplishing this is to present pairs of ordered pairs of elements of V, (a, b) and (c, d), to the Decision Maker and have him or her order them so that "the difference in value (to the Decision Maker) of a and b is less than or the same as the

functions. SEU does not address the theoretical relationship between $\mathbb{P}$ and $\mathbb{B}$ or between $\mathcal{U}$ and $\mathcal{V}$.

The simplest empirical tests in the literature of SEU involve gambles of the form $(a, A; 0, X - A)$, receiving something of positive value if A occurs and receiving nothing if $X - A$ occurs. Narens (1976) provided the following qualitative axiomatization of this situation.

Let 0 be the pure outcome of *status quo,* that is, receiving nothing, and let V be a set such that each element of V is either 0 or a pure outcome of positive value. For example, V may be consist of nonnegative amounts of money in U.S. dollars, with \$0 being the status quo 0. Gambles of the form $(a, A; 0, X - A)$ are abbreviated to (a, A).

Let $\mathcal{G}$ be a set of gambles (a, A), where (a, A) may be a proper gamble (i.e., $A \neq$ the sure event, X) or a sure gamble (i.e., $A = X$). Let $\precsim$ be a binary relation on $\mathcal{G}$ and

$$\mathcal{E} = \{A \mid (a, A) \in \mathcal{G} \text{ for some } a \in A\}.$$

The following is a slightly modified version of Narens (1976) axiomatization of SEU for this situation.

Axiom 5.1 Weak Ordering: *$\precsim$ is a weak ordering.* □

Axiom 5.2 Independence: *(i) For all A and B in $\mathcal{E}$, if for some a in V, $(a, A) \precsim (a, B)$, then for all x in V, $(x, A) \precsim (x, B)$; and (ii) for all c and d in V, if for some E in $\mathcal{E}$, $(c, E) \precsim (d, E)$, then for all F in $\mathcal{E}$, $(c, F) \precsim (d, F)$.* □

Axiom 5.3 Trade-off: *(i) For all a and b in V, if $(a, X) \precsim (b, X)$, then for some B in $\mathcal{E}$, $(a, X) \sim (b, B)$; and (ii) for all A in $\mathcal{E}$ and all x in V, there exists y in V such that $(x, A) \sim (y, X)$.* □

Axiom 5.4 Distributivity: *For all x and y in V and all A, B, C, and D in $\mathcal{E}$, if*

$$A \cap B = C \cap D = \varnothing, \ (x, A) \precsim (y, C), \ \text{and} \ (x, B) \precsim (y, D),$$

then

$$(x, A \cup B) \sim (y, C \cup D). \quad \square$$

---

difference in value of c and d." Mathematically, if certain qualitative conditions on the ordering hold, then there exists an interval scale of functions that measures the elements of V. (See Chapter 4 of Krantz, et al. for a complete discussion of this issue.)

Define $\precsim_{\mathcal{E}}$ on $\mathcal{E}$ as follows: For all A and B in $\mathcal{E}$,

$$A \precsim_{\mathcal{E}} B \text{ iff for some } x \text{ in } V,\ (x,A) \precsim (x,B).$$

Using the Axioms of Weak Ordering and part (*ii*) of the Axiom of Independence, it is easy to verify that $\precsim_{\mathcal{E}}$ is a weak ordering on $\mathcal{E}$.

Let A, B, and C be arbitrary events of $\mathcal{E}$ such that $A \precsim_{\mathcal{E}} B$ and $A \cap C = B \cap C = \varnothing$. Let x be an arbitrary element of V. Then by the definition of $\precsim_{\mathcal{E}}$,

$$(x,A) \precsim (x,B).$$

Because $\precsim$ is a weak ordering, $(x,C) \sim (x,C)$. Thus by the Axiom of Distributivity,

$$(x, A \cup C) \precsim (x, B \cup C).$$

Then by the Axiom of Independence, $A \cup C \precsim_{\mathcal{E}} B \cup C$, that is, $\cup$-monotonicity (Definition 4.2) holds for $\langle \mathcal{E}, \precsim_{\mathcal{E}} \rangle$.

Axiom 5.5 Qualitative Probability: *$\langle \mathcal{E}, \precsim_{\mathcal{E}} \rangle$ satisfies the Luce (1967) conditions for qualitative probability (Theorem 4.3), except for $\cup$-monotonicity. ($\cup$-monotonicity is not needed, because as previously shown it is implied by Axioms 5.1 to 5.4.)* □

Theorem 5.1 *Suppose Axioms 5.1 to 5.5. Then there exist a ratio scale of functions $\mathcal{U}$ on V into $\mathbb{R}^+$ and a unique function $\mathbb{P}$ on $\mathcal{E}$ such that for all u in $\mathcal{U}$ and all (a,A) and (b,B) in $\mathcal{G}$,*

$$(a,A) \precsim (b,B) \text{ iff } u(a) \cdot \mathbb{P}(A) \leq u(b) \cdot \mathbb{P}(B).$$

Proof. Follows from Theorem 2.2 of Narens (1976). □

Tversky (1967) conducted experiments involving a preference ordering $\precsim$ over gambles of the form (a, A). He found SEU did not hold for these gambles. Although he did not test directly for Distributivity, the results of his experiments imply failures of Distributivity. He also showed that the subjects' responses to the gambles were very structured: Besides finding support for the holding of Weak Ordering and Independence, he showed his data fit the following representation: There exist functions f from the objects of value (excluding the status quo, 0) into $\mathbb{R}^+$ and g from the set of uncertain events (including the sure event X but excluding the null event $\varnothing$) into $(0,1]$ such that $g(X) = 1$ and for all (a,A) and (b,B),

$$(a,A) \precsim (b,B) \text{ iff } f(a) \cdot g(A) \leq f(b) \cdot g(B).$$

5.3 Dual Bilinear and Rank Dependent Utility

The above example of Tversky and many other experiments in the literature have demonstrated that the SEU model fails to account for people's gambling behavior. Some in the literature view this is due to a failing of rational decision making by the participants. Others have argued that failures result because SEU is too narrow of a model of normative decision making under uncertainty. They hold that the SEU model should be extended to incorporate additional factors, for example, forms of risk that are not formulable within SEU. Thus, the latter researchers allow for the possibility of gambles having the same expectation in utility but differing in risk and a Decision Maker rationally preferring one to the other, for example, preferring the one with less risk. To accommodate such cases, various researchers have proposed extending the SEU model, some with the aim to better describe the empirical literature (e.g., the Dual Bilinear Utility Model described below), and others to account for normative factors that they perceive to be missing in the SEU model (e.g., the Rank Dependent Utility Model described below). These extensions or generalizations of SEU usually have representations with utility being interval scalable. Luce and Narens (1985) found for gambling structures generated by two outcome gambles that the assumption of interval scalability of utility by itself already enormously restricted the kind of models that researchers are likely to employ:

Convention 5.1 Throughout the rest of this section, the following conventions and definitions are observed:

• X stands for the sure event (which of course is nonempty).

• $\mathfrak{G} = \langle \mathcal{G}, \precsim \rangle$ is a structure of gambles such that each element of $\mathcal{G}$ is a two outcome proper gamble that has the form $(a, A; b, X - A)$. In the gamble $(a, A; b; X - A)$, a or b may also be gambles in $\mathcal{G}$.

• Recall V is the set of pure outcomes, that is, V is a nonempty set of objects of value that are not gambles. It is assumed that for each gamble $(a, A; b, X - A)$ in $\mathcal{G}$ based on the events A and $X - A$, there exist for each x and y in V another gamble $(x, A; y, X - A)$ in $\mathcal{G}$ based on the same events A and $X - A$.

• For gambles of the forms $(a, A; G, X - A)$ and $G = (c, C; d, X - C)$ of $\mathcal{G}$, it is assumed that $C \perp A$, $C \perp (X - A)$, $(X - C) \perp A$, and

$(X - C) \perp (X - A)$, where $\perp$ is the independence relation for events. A similar assumption regarding independence holds for gambles of $\mathcal{G}$ of the forms $(G, A; b, X - A)$ and $G = (c, C; d, X - C)$.

• It is assumed that $\mathcal{U}$ is an interval scale family of functions on $\mathcal{G} \cup V$.

• The following richness assumption about pure outcomes of gambles is made: For some $u \in \mathcal{U}$, the restriction of u to V, $u \restriction V$, is onto $\mathbb{R}$. Because $\mathcal{U}$ is an interval scale of functions, it follows that for all $w \in \mathcal{U}$, $w \restriction V$ is onto $\mathbb{R}$.

• For gambles $(a, A; b, X - A)$ of $\mathcal{G}$, if $u(a) < u(b)$,

$$u(a) < u((a, A; b, X - A)) < u(b)\,,$$

and if $u(a) = u(b)$, then $u((a, A; b, X - A)) = u(a)$.

• The following general representation of Luce and Narens (1985) for $\mathcal{U}$ and $\mathfrak{G}$ is assumed: For all u in $\mathcal{U}$ and all $(a, A; b, X - A)$ and $(c, C; d, X - C)$ in $\mathcal{G}$,

$$u((a, A; b, X - A)) = F(u(a), u(b), H(u(a), u(b)), A, X - A))\,, \tag{5.2}$$

and

$$\begin{aligned} &(a, A; b, X - A) \precsim (c, C; d, X - C) \\ &\text{iff } u((a, A; b, X - A)) \leq u((c, C; d, X - C))\,, \end{aligned} \tag{5.3}$$

where for fixed A, H is continuous function in terms of $u(a)$ and $u(b)$ that is into $\mathbb{R}$ and F is continuous in terms of H, $u(a)$, and $u(b)$. $\square$

Luce and Narens (1985) proved the following theorem, using slightly different assumptions and a different proof.

Theorem 5.2 *There exist u in $\mathcal{U}$ and functions $\mathbb{W}^-$ and $\mathbb{W}^+$ on the set of events occurring in the gambles in $\mathcal{G}$ into the open interval $(0,1)$ of the reals such that for each gamble $(a, A; b, X - A)$ in $\mathcal{G}$,*

$$u((a, A; b, X - A)) = \begin{cases} u(a)\mathbb{W}^-(A) + u(b)[1 - \mathbb{W}^-(A)] & \textit{if } u(a) \leq u(b) \\ u(b) & \textit{if } u(a) = u(b) \\ u(a)\mathbb{W}^+(A) + u(b)[1 - \mathbb{W}^+(A)] & \textit{if } u(a) \geq u(b)\,. \end{cases} \tag{5.4}$$

Proof. In this theorem X is a constant event. Therefore, $X - A$ varies only with A, and thus $H(u(a), u(b), A, X - A)$ only varies with $u(a)$, $u(b)$, and A. Therefore we will abuse notation a little and write, without loss of generality, "$H(u(a), u(b), A)$" for "$H(u(a), u(b), A, X - A)$" to better bring into focus the variables involved.

Because $\mathcal{U}$ is an interval scale and Equation 5.2 holds for all u in $\mathcal{U}$, it follows that for each s in $\mathbb{R}$,

$$u((a, A; b, X - A)) - s = F(u(a) - s, u(b) - s, H(u(a) - s, u(b) - s), A) .$$

Letting $s = u(b)$, the above equation becomes,

$$u((a, A; , b, X - A)) - u(b) = F(u(a) - u(b), 0, H(u(a) - u(b), 0), A) . \quad (5.5)$$

Let

$$T[u(a) - u(b), A] = F(u(a) - u(b), 0, H(u(a) - u(b), 0), A) .$$

Then Equation 5.5 becomes,

$$u((a, A; b, X - A)) = T[u(a) - u(b), A] + u(b) . \quad (5.6)$$

Because $\mathcal{U}$ is an interval scale and Equation 5.2 holds for all u in $\mathcal{U}$, it follows that for each r in $\mathbb{R}^+$,

$$ru((a, A; b, X - A)) = T[ru(a) - ru(b), A] + ru(b) ,$$

which by Equation 5.6 yields,

$$r(T[u(a) - u(b), A) + u(b)] = T[ru(a) - ru(b), A] + ru(b) ,$$

which through simplification and letting $z = u(a) - u(b)$, yields

$$rT(z, A) = T(rz, A) . \quad (5.7)$$

Because for all e and f in V, $(e, A; f, X - A) \in \mathcal{G}$ and $u \restriction V = \mathbb{R}$, it follows that z in Equation 5.7 takes on all real values. Because by hypothesis F and H are continuous, it follows that T is continuous. Equation 5.7 is a well-known functional equation whose solutions for continuous T and fixed A are as follows: For some s and t in $\mathbb{R}^+$,

$$T(z, A) = \begin{cases} s \cdot z & \text{if } z < 0 \\ 0 & \text{if } z = 0 \\ t \cdot z & \text{if } z > 0 . \end{cases}$$

Replacing $T(z, A)$ by $u(a, A; b, X - A) - u(b)$, z by $u(a) - u(b)$, and letting $s = \mathbb{W}^-(A)$ and $t = \mathbb{W}^+(A)$, we obtain,

$$u(a, A; b, X - A) - u(b) = \begin{cases} \mathbb{W}^-(A) \cdot [u(a) - u(b)] & \text{if } u(a) < u(b) \\ 0 & \text{if } u(a) = u(b) \\ \mathbb{W}^+(A) \cdot [u(a) - u(b)] & \text{if } u(a) > u(b) . \end{cases}$$

Adding $u(b)$ to the left side of the equation and adding

$\mathbb{W}^-(A)u(b) + (1 - \mathbb{W}^-(A))u(b)$ to the upper right side,

$u(b)$ to the middle right side, and

$\mathbb{W}^+(A)u(b) + (1 - \mathbb{W}^+(A))u(b)$ to the lower right side,

and simplifying and rearranging yields

$$u((a, A; b, X - A)) = \begin{cases} u(a)\mathbb{W}^-(A) + u(b)[1 - W^-(A)] & \text{if } u(a) < u(b) \\ u(b) & \text{if } u(a) = u(b) \\ u(a)\mathbb{W}^+(A) + u(b)[1 - \mathbb{W}^+(A)] & \text{if } u(a) > u(b) . \end{cases} \tag{5.8}$$

By assumption, for all $u((a, A; b, X - A))$ in $\mathcal{G}$, if $u(a) < u(b)$, then

$$u(a) < u((a, A; b, X - A)) < u(b) . \tag{5.9}$$

Taking $u(b)$ sufficiently close to $u(a)$, it is easy to show using Equations 5.8 and 5.9 that $\mathbb{W}^-(A)$, $1 - \mathbb{W}^-(A)$, $\mathbb{W}^+(A)$, and $1 - \mathbb{W}^+(A)$ are strictly between 0 and 1. □

Definition 5.4 The model expressed in Equation 5.4 is called *Dual Bilinear Utility* or *DBU* for short. When the gambles are restricted to simple gambles (i.e., gambles with pure outcomes) it is also called the *Rank Dependent Utility* or *RDU model for two outcome gambles.* □

Rank Dependent Utility for two or more outcomes have been proposed by some in the economic literature as a model of decision making under uncertainty that captures features of human economic decision behavior better than SEU.

More than two pure outcomes can appear in parts of a compound gamble in $\mathcal{G}$, for example, if a, b, and c are pure outcomes, then the gamble $G = (a, A; (c, C; d, D), B)$ has three pure outcomes appearing in parts of G. In terms of obtaining objects of value through gambling,

$$G = (a, A; (c, C; d, D), B) \text{ and } H = (a, A; c, C \cap B; d, D \cap B)$$

produce identical results. However, H is not in $\mathcal{G}$, and therefore is not formally part of Dual Binlinear Utility Theory. It is, however, part of Rank Dependent Utility Theory for three outcome gambles.

The rank dependent model for n pure outcomes states that the utility of the gamble is a weighted average of the utilities of its n outcomes. The weights are non-negative and sum to 1. They are determined by the n disjoint events that determine the gamble's outcomes together with the ordering of the outcomes. Thus for the two pure outcome gamble $G = (a, A; b, X - A)$ there are $2! = 2$ assignments of weights to A and $X - A$ depending on whether $u(a) \leq u(b)$ or $u(b) \leq u(a)$, which in our previous notation we designated as $\mathbb{W}^-(A)$ and $1 - \mathbb{W}^-(A)$ for A and $X - A$ when $u(a) \leq u(b)$, and $\mathbb{W}^+(A)$ and $1 - \mathbb{W}^+(A)$ for A and $X - A$ when $u(b) \leq u(a)$. For the case of n distinct pure outcomes there are $n!$ distinct orderings of these outcomes. Formally the rank dependent model for gambles with 3 pure outcomes is described as follows:

Definition 5.5 The *Rank Dependent Utility Model for three outcomes* describes the utility u for elements of set $\mathcal{T}$ of gambles. It is assumed that each G in $\mathcal{T}$ has the form $(d, D; e, E; f, F)$ for some partition (D, E, F) of the sure event and for some pure outcomes d, e, and f. For each gamble or object of value x, let

$$\dot{x} = u(x)\,.$$

It is further assumed that for each gamble $G = (a, A; b, B; c, C)$ in $\mathcal{T}$,

$$\dot{G} = \begin{cases} \dot{a} \cdot s_1(A,B,C) + \dot{b} \cdot [t_1(A,B,C)] + \dot{c} \cdot [1 - s_1 - t_1] & \text{if } \dot{a} \leq \dot{b} \leq \dot{c} \\ \dot{a} \cdot s_2(A,B,C) + \dot{b} \cdot [t_2(A,B,C)] + \dot{c} \cdot [1 - s_2 - t_2] & \text{if } \dot{a} \leq \dot{c} \leq \dot{b} \\ \dot{a} \cdot s_3(A,B,C) + \dot{b} \cdot [t_3(A,B,C)] + \dot{c} \cdot [1 - s_3 - t_3] & \text{if } \dot{b} \leq \dot{a} \leq \dot{c} \\ \dot{a} \cdot s_4(A,B,C) + \dot{b} \cdot [t_4(A,B,C)] + \dot{c} \cdot [1 - s_4 - t_4] & \text{if } \dot{b} \leq \dot{c} \leq \dot{a} \\ \dot{a} \cdot s_5(A,B,C) + \dot{b} \cdot [t_5(A,B,C)] + \dot{c} \cdot [1 - s_5 - t_5] & \text{if } \dot{c} \leq \dot{a} \leq \dot{b} \\ \dot{a} \cdot s_6(A,B,C) + \dot{b} \cdot [t_6(A,B,C)] + \dot{c} \cdot [1 - s_6 - t_6] & \text{if } \dot{c} \leq \dot{b} \leq \dot{a}, \end{cases}$$

where for $i = 1, \ldots, 6$, $0 < s_i + t_i < 1$. □

Consider the following two gambles in $\mathcal{G}$, where a, b, and c are pure outcomes:

(1) $G_1 = ((a, A; b, X - A), C; c, X - C)$.

(2) $G_2 = ((a, C; c, X - C), A; (b, C; c, X - C), X - A)$.

In terms of obtaining objects of value through gambling, G_1 and G_2 produce identical results, in fact, results that are identical to the simple three outcome gamble,

$$(a, A \cap C; b, (X - A) \cap C; c, X - C)\,.$$

However, results of Luce and Narens (1985) show that *the assumption of* $u(G_1) = u(G_2)$ *(for all pure outcomes a, b, and c) implies* $\mathbb{W}^+ = \mathbb{W}^-$ *in Equation 5.8, thus implying that the gambling behavior with respect to* G_1 *and* G_2 *is consistent with SEU.* Thus for Dual Bilinear Utility to be an interesting generalization of SEU, it must be the case that $u(G_1) \neq u(G_2)$ for some pure outcomes a, b, and c.

The nonequality $u(G_1) \neq u(G_2)$ just discussed is a violation of rationality. In psychology it is called a *framing effect* (Tversky and Kahneman, 1974; Kahneman and Tversky, 1979). Framing is an important psychological phenomenon, and DBU is one method of systematically modeling framing effects based on two outcome gambles. However, from the point of view of economic theory, framing effects are unwelcome and are to be excluded. RDU accomplishes this for framing effects like $u(G_1) \neq u(G_2)$ above by having its theory apply to simple gambles.[2]

Outside of employing a more general weighted average model for computing utilities of gambles, the main difference between SEU and the models given by DBU and RDU is that in SEU utilities of the objects of value are weighted by *probabilities* of their chances of occurrence, whereas in DBU and RDU they are weighted by *weights* that sum to 1. Thus DBU and RDU cannot employ the strategy of inferring subjective probabilities for events by identifying them with the weights obtained in their gambling models. To infer such probabilities, they need to be augmented by accounting for how weights are related to subjective probability. However, as discussed earlier, the same can be said for SEU: SEU provides a theory of weights where the

[2]If RDU were to extend its theory to compound gambles, it would take a route not implied by Luce and Narens' Equation 5.2, because the following is implied by that equation:

Gamble Substitutions For Pure Outcomes: For all gambles $(a, A; b, X - A)$ and $(c, C; d, X - C)$, if a, b, c, and d pure outcomes and $u(a) = u((c, C; d, X - C))$, then

$$u((a, A; b, X - A)) = u((c, C; d, X - C), A; b, X - A)\,. \tag{5.10}$$

The compound gamble $((c, C; d, X - C), A; b, X - A)$ decomposes into the simple gamble,

$$G = (c, C \cap A; d, (X - C) \cap A; b, X - A)\,.$$

Suppose $u(a) < u(b)$ and $u(c) < u(d)$. Then it follows from $u(a) = u((c, C; d, X - C))$ and RDU that $u(c) < u(a) < u(d)$. Thus, in RDU, the value of $u(G)$ in general depends on whether $u(b) < u(d)$ or $u(d) < u(b)$, but the values of $u((a, A; b, X - A))$ and $u((c, C; d, X - C))$ do not depend on this consideration. From this it follows that the following should not be a theorem of RDU: For all simple gambles $(a, A; b, X - A)$ and $(c, C; d, X - C)$, if $u(a) = u((c, C; d, X - C))$, then

$$u((a, A; b, X - A)) = u((c, C \cap A; d, (X - C) \cap A; b, X - A))\,.$$

weights are obtained through a probability function, but the theory needs to be augmented to argue that this probability function is the Decision Maker's subjective probability function for evaluating uncertainty.

SEU, DBU, and RDU are interval scalable utility theories and as such cannot distinguish between positive and negative objects of value. This is because the concepts of "0 value, positive value, and negative value" depend on a particular representative u of the underlying utility family of functions, and these concepts could change meaning if u were replaced by a different representative. In order to preserve interval scalability for this kind of situation, some theorist look at 0 value of a utility function as a "status quo" and a positive outcome, say receiving \$100, as a positive increase from the status quo, and similarly for negative outcomes. However, the real question is, "Is this how people are looking at the situation?" I think not. I believe it is best when dealing with questions of scale type to provide a qualitative model involving all the relevant considerations, and then see what scale type the qualitative model implies. In this case, a model of the form

$$\mathfrak{G} = \langle \mathcal{G} \cup V, \precsim, V, \mathbf{0}, P, N \rangle$$

appears to me to capture the relevant structure when subjects are presented gambles with positive and negative outcomes, where $\langle \mathcal{G}, \precsim \rangle$ is a gambling structure with V the set of objects of value, $\mathbf{0}$ is the status quo, P is the set of positive elements of V, and N the set of negative elements of V. Such situations, when appropriately axiomatized, yield families of utility functions $\mathcal{U}$ that are ratio scales such that for each u in $\mathcal{U}$, p in P, and n in N, $u(\mathbf{0}) = 0$, $u(p) > 0$, and $u(n) < 0$. In the model $\mathfrak{G}$, it is assumed that the subject is evaluating the object $\mathbf{0}$ as having no value, evaluating objects in P as having positive values, and objects in N as having negative values. If one were to assume interval scalability, then $\mathfrak{G}$ would not be the correct formulation of the situation: The correct one for that case would have $\mathbf{0}$, P, and N subscripted by elements of V, that is, be the qualitative model,

$$\mathfrak{H} = \langle \mathcal{G} \cup V, \precsim, V, \mathbf{0}_a, P_a, N_a \rangle_{a \in V} .$$

In $\mathfrak{H}$, the status quo can change, and thus the object of value \$100 is not always viewable as being positive, for example, when $a = \$100$.

For understanding human gambling behavior in structurally richer situations than are traditionally part of economic utility theory, for example, in gambling situations where outcomes are *sets* of objects of value as well as *singleton* objects of value, ratio scalable models are usually required (e.g., see Luce, 2000). Because such richer situations occur normatively, it is therefore not necessary normatively that utility is measured on an interval scale. Thus social, economic, and philosophical theories and arguments that employ in an important way a normative assumption of interval scalable utility should be viewed with some suspicion.

Chapter 6

Axioms for Choice Proportions

6.1 Brief Overview

This chapter provides a qualitative axiomatization of a special case of conditional probability. The axiomatization is generalized in Chapter 7 so that it applies to belief situations where the belief function has two dimensions—a probabilistic dimension measured by a probability function, and another dimension measured by a different function.

The qualitative axiomatization presented in the chapter has a representation theorem that corresponds to an important choice model introduced by Luce (1959). Luce's model is characterized by a quantitative condition known today as *Luce's Choice Axiom.* One of the many interpretations of Luce's axiom is that it describes a situation of conditional probability.[1] Thus a qualitative axiomatization of Luce's Choice Axiom can also be viewed as a qualitative axiomatization of conditional probability.

The axiomatization presented in this chapter is designed to isolate a key property that fails for the theories of belief describe later in Chapters 7 and 9. The axiomatization minus this key property produces a generalization of conditional probability that apply to a descriptive theory of human judgments of probability (Chapter 7) and a normative theory for the prob-

[1]Luce formulated his axiom for a boolean algebra of events $\mathcal{E}$ on a finite set X. His axiom also applies to infinite situations where events do not form a boolean algebra but conditional probabilities nevertheless exist. In some of these situations, it is impossible to find an unconditional probability function that gives rise to the conditional probabilities. An example is when the conditional probabilities $\mathbb{P}(A|B)$ for finite subsets A and B, $B \neq \varnothing$, of an infinite set X is defined by,

$$\mathbb{P}(A|B) = \frac{\text{number of elements in } A \cap B}{\text{number of elements in } B}\,.$$

abilities of scientific refutations (Chapter 9).

6.2 Luce's Choice Property

Convention 6.1 The theory of conditional probability presented in this chapter is founded on ideas from choice theory. There $A \mid B$ is interpreted as the event of choosing an element of the set A when the set B is presented. In this situation, A is often called the *choice set, choice,* or *focus (of* $(A \mid B)$*),* and B the *context (of* $(A \mid B)$*).* Usually the focus is assumed to be a subset of the context. Also in choice theory the notation "P_B" is used to say that B is a nonempty set and P_B is a finitely additive probability function on the boolean algebra of all *nonempty* subsets of B. These conventions from choice theory are used in various parts of the book. □

Definition 6.1 Let Y be a nonempty finite set. Then the set of probability functions

$$\{P_C \mid C \text{ is a nonempty subset of } Y\}$$

is said to satisfy *Luce's Choice Axiom* if and only if for all nonempty subsets A and B of Y such that $A \subseteq B$,

$$\frac{P_Y(A)}{P_Y(B)} = P_B(A). \quad \square \tag{6.1}$$

Luce (1959) showed the following theorem:

Theorem 6.1 *Let Y be a nonempty finite set and*

$$\{P_C \mid C \textit{ is a nonempty subset of } Y\}$$

be a system of probability functions satisfying Luce's Choice Axiom (Definition 6.1). Let $\mathcal{U}$ be the set of all functions u such that for all nonempty subsets A and B of Y,

$$P_B(A) = \frac{\sum_{a \in A} u(a)}{\sum_{b \in B} u(b)}. \tag{6.2}$$

Then $\mathcal{U}$ is a ratio scale.

Proof. For each y in Y, let $u(y) = P_Y(\{y\})$. Then by Equation 6.1,

$$P_B(A) = \frac{P_Y(A)}{P_Y(B)} = \frac{\frac{\sum_{a \in A} \mathbb{P}_Y(\{a\})}{\sum_{y \in Y} \mathbb{P}_Y(\{y\})}}{\frac{\sum_{b \in B} \mathbb{P}_Y(\{b\})}{\sum_{y \in Y} \mathbb{P}_Y(\{y\})}} = \frac{\sum_{y \in A} u(y)}{\sum_{y \in B} u(y)}. \tag{6.3}$$

Then $\mathcal{U} \neq \varnothing$, because $u \in \mathcal{U}$. Let $\mathcal{U} = \{ru \,|\, r \in \mathbb{R}^+\}$. Then for each t in $\mathcal{U}$, if t replaces u in Equation 6.3, then t satisfies Equation 6.3. Suppose s is a function that satisfies Equation 6.3 with s in place of u. It only needs to be shown that $rs \in \mathcal{U}$ for some r in $\mathbb{R}^+$. Then for $B = Y$ in Equation 6.3 and $A = \{y\}$ for an arbitrary y in Y,

$$\frac{s(y)}{\sum_{x \in Y} s(x)} = \frac{P_Y(\{y\})}{P_Y(Y)} = \frac{u(y)}{\sum_{x \in Y} u(x)},$$

and thus for each y in Y, $s(y) = ru(y)$ for

$$r = \frac{\sum_{x \in Y} s(x)}{\sum_{x \in Y} u(x)}. \quad \square$$

Although Theorem 6.1 is straightforward with a simple proof, its application in substantive domains of science has produced deep and insightful results by providing interesting interpretations for the elements u in $\mathcal{U}$. In this book, u is interpreted as a function that measures the support for a proposition based on evidence. This interpretation is generalized in Chapters 7 and 10 to apply to psychological judgments of probability.[2]

[2] Another interesting interpretation of u is obtained by generalizing Theorem 6.1 to the infinite case described in the following theorem. (The proof of the theorem is not presented here. It can be shown using methods of nonstandard analysis developed in Robinson, 1966.)

Theorem Suppose X is an infinite set and for each finite nonempty subset Y of X, $\{P_B \,|\, B \text{ is a nonempty subset of } Y\}$ is a system of probabilities that satisfies Luce's Choice Axiom. Then the following two statements hold.

1. There exists a function u from X into $\mathbb{R}^+$ such that for each nonempty finite subset B of X and each nonempty subset A of B,

$$P_B(A) = \frac{\sum_{a \in A} u(a)}{\sum_{b \in B} u(b)}. \tag{6.4}$$

2. Let u be as in Equation 6.4. Suppose for each positive real number r there exists a finite subset F of X such that

$$\sum_{a \in F} u(a) > r.$$

Then there exist a boolean algebra $\mathcal{E}$ of subsets of a set ${}^\star X$, a totally ordered field extension of the reals ${}^\star\mathfrak{R} = \langle {}^\star\mathbb{R}, \leq, +, \cdot, 0, 1\rangle$, a positive infinitesimal element α of ${}^\star\mathfrak{R}$, and a function $\mathbb{P}$ from $\mathcal{E}$ into ${}^\star[0,1]$ such that

(*i*) $\mathbb{P}({}^\star X) = 1$, $\mathbb{P}(\varnothing) = 0$, and for all A and B in $\mathcal{E}$ such that $A \cap B = \varnothing$,

$$\mathbb{P}(A \cup B) = \mathbb{P}(A) + \mathbb{P}(B),$$

and

6.3 BTL Model of Choice

Much behavioral science research involves the modeling of the probabilistic choice of objects from a set of alternatives. A particularly important model is one in which objects are assigned positive numbers by a function u so that the probability p that object a is chosen from the set of alternatives $\{a \mid a_1, \ldots, a_n\}$ is given by the equation,

$$p = \frac{u(a)}{u(a) + u(a_1) + \cdots + u(a_n)}. \tag{6.5}$$

In the literature, this model is often called the *Bradley-Terry-Luce Model,* which is often abbreviated to the *BTL Model.* A special case of it,

$$p = \frac{u(a)}{u(a) + u(a_1)},$$

was used by the famous set-theorist E. Zermelo to describe the power of chess players (Zermelo, 1929). (In current chess usage, $u(a)$ would be a "rating" that stands for player a's ability to win chess games, and

$$p = \frac{u(a)}{u(a) + u(a_1)},$$

is the probability that player a would beat player a_1 in a chess match. The rating system used by chess organizations differ from Zermelo's but produces approximately the same probability estimates as Zermelo's, e.g., see Batchelder and Bershad, 1979). The choice models implicit in Equation 6.5 have also been used by Bradley and Terry (1952), Luce (1959), and many others in behavioral applications.

6.4 Basic Choice Axioms

This section presents a qualitative axiomatization of the BTL model due to Narens (2003). This axiomatization was not only designed to qual-

---

(*ii*) for each finite nonempty subset A of X, $\mathbb{P}(A) = \alpha \cdot u(A)$, where

$$u(A) = \sum_{a \in A} u(a). \quad \square$$

The previous theorem shows that when $u(A)$ takes on arbitrary large values for some finite subsets A of X, then u can be interpreted as corresponding to a portion of the infinitesimal part of an unconditional probability function taking values in a totally ordered field extension of the reals.

itatively capture the essential features of the BTL model, but was also designed to be generalizable to account for an interesting class of probabilistic judgments by deleting a specific axiom. The generalization is presented in Chapter 7.

Convention 6.2 Throughout this section, X will denote an infinite set of objects and $\precsim$ a binary relation on X. □

Convention 6.3 By definition, a *context (of X)* is a nonempty finite subset of X. Throughout the remainder of the chapter it is assumed that all contexts have at least two elements. $\mathcal{C}$ will denote the set of contexts.

The notation $(a \mid C)$ stands for "C is a context and $a \in C$." "$(a \mid C)$" is often read as "strength of a in the context C." When $C = \{a, a_1, \ldots, a_n\}$, $(a \mid C)$ is often written as $(a \mid a, a_1, \ldots, a_n)$. By convention, the notation $(a \mid a, a_1, \ldots, a_n)$ assumes the elements $a, a_1, \ldots, a_n$ are distinct. □

Definition 6.2 The *Basic Choice Axioms* consist of the following 14 axioms. □

Axiom 6.1 *$\precsim$ is a weak ordering on the set*

$$\{(a, C) \mid C \in \mathcal{C} \text{ and } a \in C\}. \quad \square$$

Axiom 6.2 *Suppose $A \in \mathcal{C}$, $a \in A$, and B is a nonempty finite subset of X such that $B \cap A = \varnothing$. Then $(a \mid A \cup B) \prec (a \mid A)$.* □

Axiom 6.3 *Suppose A and B are in $\mathcal{C}$ and C is a nonempty finite subset of X such that $A \cap C = B \cap C = \varnothing$, and suppose a and b are elements of X. Then*

(i) if $a \in A$ and $a \in B$, then

$$(a \mid A) \precsim (a \mid B) \text{ iff } (a \mid A \cup C) \precsim (a \mid B \cup C),$$

(ii) if $a \in A$ and $b \in A$, then

$$(a \mid A) \precsim (b \mid A) \text{ iff } (a \mid A \cup C) \precsim (b \mid A \cup C),$$

and

(iii) if $a \in A$, $a \in B$, $b \in A$, and $b \in B$, then

$$(a \mid A) \precsim (a \mid B) \text{ iff } (b \mid A) \precsim (b \mid B). \quad \square$$

Through the use of Axiom 6.3, $\precsim$ induces natural orderings on X and $\mathcal{C}$ as follows:

Definition 6.3 By definition $\precsim_X$ is the binary relation on X such that for all a and b in X, $a \precsim_X b$ if and only if there exists a finite set C such that

$$C \subseteq X - \{a,b\} \text{ and } (a \mid \{a,b\} \cup C) \precsim (b \mid \{a,b\} \cup C). \quad \square$$

Note that it follows from Axiom 6.3(ii) and Definition 6.3 above that $a \precsim_X b$ if and only if *for all* C,

$$\text{if } C \text{ is finite and } C \subseteq X - \{a,b\} \text{ then } (a \mid \{a,b\} \cup C) \precsim (b \mid \{a,b\} \cup C).$$

Definition 6.4 By definition, $\precsim_{\mathcal{C}}$ is the binary relation on $\mathcal{C}$ such that for all C and D in $\mathcal{C}$, $D \precsim_{\mathcal{C}} C$ if and only if for some a in X such that a is not in $C \cup D$,

$$(a \mid \{a\} \cup C) \precsim (a \mid \{a\} \cup D). \quad \square$$

Lemma 6.1 *$D \precsim_{\mathcal{C}} C$ if and only if* for all *b in X such that b is not in $C \cup D$*

$$(b \mid \{b\} \cup C) \precsim (b \mid \{b\} \cup D).$$

Proof. Suppose a and b are arbitrary elements of X, D and C are in $\mathcal{C}$ and $\{a,b\} \cap (D \cup C) = \varnothing$. Also suppose a is such that

$$(a \mid \{a\} \cup C) \precsim (a \mid \{a\} \cup D).$$

Then by Axiom 6.3(i),

$$(a \mid \{a,b\} \cup C) \precsim (a \mid \{a,b\} \cup D).$$

By Axiom 6.3(iii),

$$(b \mid \{a,b\} \cup C) \precsim (b \mid \{a,b\} \cup D),$$

which by Axiom 6.3(i) yields,

$$(b \mid \{b\} \cup C) \precsim (b \mid \{b\} \cup D). \quad \square$$

Definition 6.4 says that $D \precsim_{\mathcal{C}} C$ if and only if for some a in $X - (C \cup D)$ the strength of a in context $\{a\} \cup C$ is *less than or the same* as its strength in context $\{a\} \cup D$. It immediately follows from the fact that $\precsim$ is a weak ordering that the induced orderings on X and $\mathcal{C}$ described above are also weak orderings.

Definition 6.5 A function u is said to be a *basic choice representation* for $\precsim$ if and only if the following four conditions hold:

1. u is a function from X into $\mathbb{R}^+$.

2. For all a and b in X, $a \precsim_X b$ iff $u(a) \leq u(b)$.

3. For all C and D in $\mathcal{C}$,

$$C \precsim_{\mathcal{C}} D \text{ iff } \sum_{e \in C} u(e) \leq \sum_{e \in D} u(e).$$

4. For all distinct $a, a_1, \ldots, a_n$ and all distinct $b, b_1, \ldots, b_m$,

$$(a \mid a, a_1, \ldots, a_n) \precsim (b \mid b, b_1, \ldots, b_m)$$

   if and only if

$$\frac{u(a)}{u(a) + u(a_1) + \cdots + u(a_n)} \leq \frac{u(b)}{u(b) + u(b_1) + \cdots + u(b_m)}. \quad \square$$

It is easy to verify directly through Definition 6.5 that Axioms 6.2 and 6.3 are necessary conditions for the existence of a basic choice representation for $\precsim$. Similarly, direct verification shows that the following three axioms are also necessary for the existence of a basic choice representation for $\precsim$:

Axiom 6.4 *For all a, b, c, e, and f in X, if*

$$a \neq b,\ a \neq c,\ (e \mid e, a) \sim (e \mid e, b),\ \textit{and}\ (f \mid f, a) \sim (f \mid f, c),$$

then $\{a, b\} \sim_{\mathcal{C}} \{a, c\}$. $\square$

Axiom 6.5 *For all a, a', b, and b' in X and all A, A', B, and B' in $\mathcal{C}$, if $a \sim_X a'$, $b \sim_X b'$, and $A \cap A' = B \cap B' = \varnothing$, then the following two statements are true:*

1. *If $(a \mid A) \precsim (b \mid B)$ and $(a' \mid A') \precsim (b' \mid B')$, then*

$$(a \mid A \cup A') \precsim (b \mid B \cup B').$$

2. *If $(a \mid A) \prec (b \mid B)$ and $(a' \mid A') \precsim (b' \mid B')$, then*

$$(a \mid A \cup A') \prec (b \mid B \cup B'). \quad \square$$

Axiom 6.6 *Suppose a and b are arbitrary elements of X, A and B are arbitrary elements of $\mathcal{C}$ and $a \in A$, and $b \in B$. Then the following two statements are true:*

1. *If $A \sim_{\mathcal{C}} B$ then: $a \precsim_X b$ iff $(a \mid A) \precsim (b \mid B)$.*
2. *If $a \sim_X b$ then: $A \precsim_{\mathcal{C}} B$ iff $(b \mid B) \precsim (a \mid A)$.* □

Axiom 6.5 in the presence of the other axioms corresponds to a well-investigated axiom of measurement theory called "distributivity". Axiom 6.6 corresponds to another well-investigated axiom of measurement theory called "monotonicity."

To obtain strong results about basic belief representations for $\precsim$, additional axioms are needed. The following six show that the situation under consideration is rich in objects and contexts.

Axiom 6.7 *For all $a \in X$ and $B \in \mathcal{C}$, if $A \precsim_{\mathcal{C}} B$ for some A in $\mathcal{C}$ such that $a \in A$, then there exist $c \in X$ and $C \in \mathcal{C}$ such that $a \sim_X c$, $B \sim_{\mathcal{C}} C$, and $c \in C$.* □

Axiom 6.8 *For all a and b in X and all A in $\mathcal{C}$, if $a \in A$ and $b \precsim_X a$, then there exists c in X and C in $\mathcal{C}$ such that $b \sim_X c$, $A \sim_{\mathcal{C}} C$, and $c \in C$.* □

Axiom 6.9 *For all A and B in $\mathcal{C}$ and all $b \in B$, there exist $c \in X$ and C in $\mathcal{C}$ such that $b \sim_X c$, $B \sim_{\mathcal{C}} C$, $A \cap C = \varnothing$ and $(b \mid B) \sim (c \mid C)$.* □

Axiom 6.10 *For each $A = \{a_1, \ldots, a_n\}$ in $\mathcal{C}$ there exist $A' = \{a'_1, \ldots, a'_n\}$ in $\mathcal{C}$ and e in X such that $A \cap A' = \varnothing$, $\{e\} \cap (A \cup A') = \varnothing$, and for $i = 1, \ldots, n$,*

$$(e \mid e, a_i) \sim (e \mid e, a'_i). \quad \square$$

Axiom 6.11 *For all A and B in $\mathcal{C}$, if $A \prec_{\mathcal{C}} B$ then there exists C in $\mathcal{C}$ such that $A \cap C = \varnothing$ and $A \cup C \prec_{\mathcal{C}} B$.* □

Axiom 6.12 *The following two statements are true:*

1. *For all a and b in X and all A in $\mathcal{C}$, if $a \in A$ and $a \precsim_X b$, then there exist c and C such that $c \in C$ and*
$$c \sim_X b \text{ and } (c \mid C) \sim (a \mid A).$$
2. *For all a in X and all A and B in $\mathcal{C}$, if $a \in A$, then there exist c and C such that*
$$C \sim_{\mathcal{C}} B \text{ and } (c \mid C) \sim (a \mid A). \quad \square$$

The next two axioms are necessary for the existence of a basic choice representation.

Axiom 6.13 (Archimedean Axiom) *For all* $A, B, B_1, \ldots, B_i, \ldots$ *in* $\mathcal{C}$, *if for all distinct* i *and* j *in* $\mathbb{I}^+$ $B_i \cap B_j = \varnothing$ *and* $B_i \sim_{\mathcal{C}} B$, *then for some* $n \in \mathbb{I}^+$,

$$A \prec_{\mathcal{C}} \bigcup_{k=1}^{n} B_k \,. \quad \square$$

Axiom 6.14 (Binary Symmetry) *Let* a, b, c, *and* d *be arbitrary, distinct elements of* X *such that*

$$(a \mid a, b) \sim (b \mid a, b) \text{ and } (c \mid c, d) \sim (d \mid c, d)\,.$$

Then

$$(a \mid a, b) \sim (c \mid c, d)$$

and

$$(a \mid a, c) \sim (b \mid b, d). \quad \square$$

In the context of conditional probability, Binary Symmetry asserts that if a and b have equal likelihood of occurring and c and d have equal likelihood of occurring, then the conditional probabilities of $(a \mid a, b)$ and $(c \mid c, d)$ are $\frac{1}{2}$ (and thus are the "same"), and the conditional probabilities of $(a \mid a, c)$ and $(b \mid b, d)$ are the same.

Theorem 6.2 *Assume the Basic Choice Axioms, that is, assume Axioms 6.1 to 6.14. Then the following two statements hold:*

1. (Representation Theorem) *There exists a basic choice representation for* $\precsim$ *(Definition 6.5).*

2. (Uniqueness Theorem) *The set of basic choice representations for* $\precsim$ *is a ratio scale (Definition 1.11).*

Proof. Theorem 2.2 of Narens (2003). $\square$

One interpretation of the Representation Theorem part of Theorem 6.2 is that the Basic Choice Axioms qualitatively describe a situation of conditional probability.

For intuitive purposes and for exposition, Axioms 6.1 to 6.14 are divided into three rough categories: (*i*) *substantive axioms* that reveal important structural relationships about conditional probability; (*ii*) *richness axioms* that guarantee that a rich probabilistic situation is under consideration; and *other axioms* that belong to neither categories (*i*) nor (*ii*).

The role of the substantive axioms is to describe a rich setting for conditional probability. Axioms 6.1 to 6.6 and Binary Symmetry (Axiom 6.14) are substantive axioms. The role of the richness and other axioms are to guarantee that such a description can be made easily and will work. Axioms 6.7 to 6.12 are examples of richness axioms. They assert the existence of certain kinds of solvability relations among events in terms of $\sim$ and $\prec$. The Archimedean axiom (Axiom 6.13) is an example of an "other axiom." Its role is to eliminate infinitesimally small events so that the real number system can be employed to measure the degrees of uncertainty implicit in the comparison of events by $\precsim$.

The Basic Choice Axioms (Definition 6.2) provide (via Theorem 6.2) a qualitative description of the BTL model of Choice. There are many ways in which the ordering $\precsim$ on choice strengths of objects in contexts are established empirically. One example is where "$(a \mid A) \precsim (b \mid B)$" stand for, "The conditional probability of a being chosen from A is less than or equal to the conditional probability of b being chosen from B, and Luce's Choice Axiom holds on the conditional probabilities generated by $\{(c|C) \mid c \in X \text{ and } C \in \mathcal{C}\}$." Other kinds of examples can be given where $\precsim$ yields the Basic Choice Axioms by different means, for example, $(a \mid A) \precsim (b \mid B)$ if and only if the reaction time in choosing a from A is at least as long as the reaction time of choosing b from B. The added flexibility of multiple interpretations of qualitative axiomatizations is one of the great strengths of the measurement-theoretic approach to mathematical modeling.

There are a number of ways of extending the Basic Choice Axioms to produce a qualitative axiomatization of a system of conditional probability (or alternatively, Luce's Choice Axiom). In the extended axiomatization, conditional probabilities are defined for all conditional events of the form $A \mid B$, A and B finite subsets of X and $B \neq \varnothing$. Chapter 7 provides such an extended axiomatization (Section 7.3). The extended axiomatization is designed so that the deletion of Binary Symmetry produces interesting and applicable models of degrees of belief.

Chapter 7

Conditional Belief

7.1 Introduction

It is intuitively compelling that beliefs can be compared in terms of strength. As discussed in Chapter 4, many in the foundations of probability believe that the rational comparing of degrees of belief necessarily correspond to comparisons of probabilities from a finitely additive or σ-additive probability function. Although other rational means of comparison have been proposed, these generally lack interesting mathematical structure and effective means for calculating and manipulating degrees of belief—features I consider to be necessary for a proposed replacement of classical probability theory. Descriptive theories of belief also generally lack similar mathematical structure.

This chapter investigates an alternative to conditional probability as a theory of conditional belief. The alternative shares many characteristics with conditional probability and has the calculative power at the same level as finitely additive conditional probability. It arises from a qualitative axiomatization of conditional probability that has a controversial feature isolated as an axiom. The axiom is deleted and the models of the remaining axioms are characterized terms of a quantitative formula. This strategy is in someways similar to the one employed by geometers near the start of the 19th century to characterize noneuclidean plane geometries. There, a controversial axiom of Euclid—the parallel postulate—was deleted to produce a geometrical system that is today called "absolute geometry." Absolute geometry has two natural axiomatic extensions: one producing euclidean geometry by reintroducing the parallel postulate, and one producing hyperbolic geometry by adding an axiom saying that for each point p not on a line l there exist infinitely many lines through p parallel to l. In this chapter, a similar axiomatic program is carried out. Its goal is to discover plausible alternatives to the Kolmogorov theory for describing degrees of belief.

It should be noted that the purpose of the axiomatizations used in this chapter is to discover quantitative models of belief that may be of importance for belief theory. With this in mind, it is not important for the "remaining axioms" to be elegant or simple or subject to test. Once the quantitative characterization of the "remaining axioms" is known, theories of belief can be studied directly in terms of it—as is done in this book—or it and specific instances of it can be used as targets for further and perhaps more elegant and better testable qualitative axiomatizations.

The alternative to Kolmogorov theory presented in this chapter arises by generalizing the Basic Choice Axioms (Definition 6.2) to produce a version of qualitative conditional probability. The generalization arises by deleting the Axiom of Binary Symmetry (Axiom 6.14) from the Basic Choice Axioms, and characterizing the models of the remaining axioms in terms of a quantitative formula.

Convention 7.1 Throughout this chapter, the notation, definitions, and concepts of Chapter 6 are assumed. □

Recall that the Axiom of Binary Symmetry asserts that for all distinct states of the world, a, b, c, and d, if

$$(a \mid a, b) \sim (b \mid a, b) \text{ and } (c \mid c, d) \sim (d|c, d)\,, \tag{7.1}$$

then

$$(a \mid a, b) \sim (c \mid c, d) \tag{7.2}$$

and

$$(a \mid a, c) \sim (b \mid b, d)\,. \tag{7.3}$$

Narens (2003), echoing longstanding concerns of others, notes the following problem with taking Binary Symmetry as a necessary condition for a *general theory* of belief:

> In the context of the other axioms for conditional probability, Binary Symmetry asserts that if a and b have equal likelihood of occurring and c and d have equal likelihood of occurring, then the conditional probabilities of $(a \mid a, b)$ and $(c \mid c, d)$ are $\frac{1}{2}$, and the conditional probabilities of $(a \mid a, c)$ and $(b \mid b, d)$ are the same.

> Suppose Equation [7.1] and the judgment of equal likelihood of the occurrences of a and b, given either a or b occurs, is based on much information about a and b and a good understanding of the nature of the uncertainty involved, and the judgment of equal likelihood of the occurrences of c and d, given either c or d occurs, is due to the lack of knowledge of c and d, for example, due to complete ignorance of c and d. Then, because of the differences in the understanding of the nature of the probabilities involve, a lower degree of belief may be assigned to $(c|c, d)$ than to $(a \mid a, b)$, thus invalidating Equation [7.2].
>
> Suppose Equation [7.1] and the judgments of the likelihoods of the occurrences of a and c, given either a or c occurs, are based on much information about a and c and the nature of the uncertainty involved, and the judgment the likelihoods of the occurrences of b and d given either b or d occurs is due to the lack of knowledge of b and d. This may result in different degrees of belief being assigned to $(a \mid a, c)$ and $(b \mid b, d)$, and such an assignment would invalidate Equation [7.3]. *(p. 2)*

An interesting consequence of the alternative theory of belief presented in this chapter is that a representation theorem shows that uncertainty is measured in terms of two functions. One of these is interpretable as a conditional probability function. The other has many interpretations that vary with the nature of the uncertainty under consideration. The alternative theory shares many properties with classical probability theory, and it has the mathematical structure and calculative richness equal to that of finitely additive probability theory.

Section 7.2 presents a generalization of the Basic Choice Axioms (Definition 6.2) by deleting the Axiom of Binary Symmetry. The resulting axiom system is called the *Basic Belief Axioms.* Both the Basic Choice Axioms and Basic Belief Axioms are about conditional choices of the form $(A \mid B)$, where B is a nonempty finite subset of the states of the world, X, and $A \subseteq B$. For both axiom systems, B is assumed to have at least two elements and $A = \{a\}$ for some a in B. In Section 7.2, the Basic Belief Axioms are extended to situations where A is an arbitrary subset of a nonempty finite subset of B. The resulting extended axiom system is called the *Belief Axioms.* The Basic Belief Axioms and the Belief Axioms have similar representation and uniqueness theorems. The following is the representation and uniqueness theorem for the Basic Belief Axioms:

Theorem 7.1 *Assume the Basic Belief Axioms. Then there exist ratio scales of functions $\mathcal{U}$ and $\mathcal{V}$ such that for all u in $\mathcal{U}$ and v in $\mathcal{V}$ and all* *conditional choices* $(a \mid A)$ *and* $(b \mid B)$,

$$(a|A) \precsim (b|B) \ \text{ iff } \ \frac{u(a)}{\sum_{a \in A} u(a)} \cdot v(a) \leq \frac{u(b)}{\sum_{b \in B} u(b)} \cdot v(b)\,. \quad \square \tag{7.4}$$

Letting

$$\mathbb{P}(a \mid A) = \frac{u(a)}{\sum_{a \in A} u(a)} \ \text{ and } \ \mathbb{P}(a \mid B) = \frac{u(b)}{\sum_{b \in B} u(b)}\,,$$

Equation 7.4 then becomes

$$(a \mid A) \precsim (b \mid B) \ \text{ iff } \ \mathbb{P}(a \mid A) \cdot v(a) \leq \mathbb{P}(b \mid B) \cdot v(b)\,. \tag{7.5}$$

One interpretation of Equation 7.5 is that the uncertainty of $(a \mid A)$ is measured in terms of a conditional probability function $\mathbb{P}$ that measures the probabilistic strength of a being chosen from A, and in terms of another function v that measures some nonprobabilistic dimension of uncertainty that in this book is called "definiteness." "Definiteness" has a number of interpretations. One is as an opposite of "ambiguity," that is, greater definiteness = lesser ambiguity and vice versa.

The Belief Axioms generalize the Basic Choice Axioms in two ways: (i) by deleting the Axiom of Binary Symmetry, and (ii) by extending the domain of conditional choices to choices of the form $(A \mid B)$ where A is an arbitrary subset of a nonempty finite subset B of the set of objects to be chosen, X. It is shown in Section 7.2 that the Belief Axioms yield the following generalization of Equation 7.5 with v being extended to all finite subsets of X: For all conditional events $(A \mid B)$ and $(C \mid D)$ of X,

$$(A \mid B) \precsim (C \mid D) \ \text{ iff } \ \frac{\sum_{a \in A} u(a)}{\sum_{b \in B} u(b)} \cdot v(A) \leq \frac{\sum_{c \in C} u(c)}{\sum_{d \in D} u(d)} \cdot v(C)\,. \tag{7.6}$$

Letting

$$\mathbb{P}(A|B) = \frac{\sum_{a \in A} u(a)}{\sum_{b \in B} u(b)} \ \text{ and } \ \mathbb{P}(C|D) = \frac{\sum_{c \in C} u(c)}{\sum_{d \in D} u(d)}\,,$$

then yields,

$$(A \mid B) \precsim (C \mid D) \ \text{ iff } \ \mathbb{P}(A|B) \cdot v(A) \leq \mathbb{P}(C|D) \cdot v(C)\,, \tag{7.7}$$

where $\mathbb{P}$ is a finitely additive conditional probability function.

In Section 7.3, Binary Symmetry is added to the Belief Axioms to produce the *Choice Axioms*. A theorem shows that the Choice Axioms hold if and only if there exists a finitely additive conditional probability function $\mathbb{P}$ such that for all conditional beliefs $(A \mid B)$ and $(C \mid D)$,

$$(A \mid B) \precsim (C \mid D) \text{ iff } \mathbb{P}(A|B) \leq \mathbb{P}(C|D).$$

A consequence of this is that if v in Equation 7.7 is a constant function, then the Choice Axioms hold.

This book provides both empirical and normative interpretations of Equations 7.6 and 7.7. The empirical interpretations concern human probability judgment and are discussed in Sections 7.4 and 7.5. A normative interpretation is given in Chapter 9.

In Section 7.4 the representation of conditional beliefs given by Equations 7.6 and 7.7 is used to explain the empirical results that occur in a much empirically and theoretically investigated situation called "Ellsberg's Paradox." Here v is used to measure an opposite of what in the Ellsberg's Paradox literature is called "ambiguity." The data giving rise to the paradox assumes only unconditional probabilities. The representation of conditional beliefs given by Equations 7.6 and 7.7 provides an explanation for the paradox. However, a thought-experiment is presented that suggests that the representation will likely fail if the supposed mechanisms giving rise to it and the Ellsberg Paradox are extended to richer data sets involving conditional rather than only unconditional probabilities.

In Section 7.5 a different interpretation to Equations 7.6 and 7.7 is used to model a descriptive theory of human probability judgments called "Support Theory." There v is used to account for differences in probability judgments of an event due to how the event is described to the subject. A different and more extensive account of human probability judgments is given in Chapter 10.

7.2 Belief Axioms

Definition 7.1 The *Basic Belief Axioms* consist of the Basic Choice Axioms (Definition 6.2) except for the Axiom of Binary Symmetry. □

Definition 7.2 An ordered pair of functions $\langle u, v \rangle$ is said to be a *basic belief representation for* $\precsim$ if and only if the following three conditions hold:

1. u and v are functions from X into $\mathbb{R}^+$.

2. For all C and D in $\mathcal{C}$,

$$C \precsim_{\mathcal{C}} D \text{ iff } \sum_{e \in C} u(e) \leq \sum_{e \in D} u(e).$$

3. For all distinct $a, a_1, \ldots, a_n$ and all distinct $b, b_1, \ldots, b_m$,

$$(a \mid a, a_1, \ldots, a_n) \precsim (b \mid b, b_1, \ldots, b_m)$$

if and only if

$$\frac{u(a)}{u(a) + u(a_1) + \cdots + u(a_n)} \cdot v(a) \leq \frac{u(b)}{u(b) + u(b_1) + \cdots + u(b_m)} \cdot v(b).$$

Theorem 7.2 *Assume the Basic Belief Axioms (Definition 7.1). Then the following two statements hold:*

1. (Representation Theorem) *There exists a basic belief representation for $\precsim$ (Definition 7.2).*
2. (Uniqueness Theorem) *Let*

$$\mathcal{U} = \{u \mid \textit{there exists } v \textit{ such that } \langle u, v \rangle \textit{ is a basic belief representation for } \precsim\}$$

and

$$\mathcal{V} = \{v \mid \textit{there exists } u \textit{ such that } \langle u, v \rangle \textit{ is a basic belief representation for } \precsim\}.$$

Then $\mathcal{U}$ and $\mathcal{V}$ are ratio scales.

Proof. Theorem 2.2 of Narens (2003). □

The following definitions and axioms extend the basic belief axioms to situations involving more general choices of the form $(A \mid B)$, where $\varnothing \subseteq A \subseteq B \subset X$ and B is finite and nonempty.

Definition 7.3 Let

$$\mathcal{F} = \{F \mid F \text{ is a finite subset of } X\}.$$

Elements of $\mathcal{F}$ are called *finite events (of X)*. Elements of $\mathcal{F} - \{\varnothing\}$ are called *context events (of X)*. In the notation "$(A \mid B)$," where A is in $\mathcal{F}$, B is in $\mathcal{F} - \{\varnothing\}$, and $A \subseteq B$, A is called the *focal event* of $(A \mid B)$ and B is called the *context* or *conditioning event* of $(A \mid B)$. By definition, the *finite conditional events (of X)* consists of all $(A \mid B)$ where A is in $\mathcal{F}$, B is in $\mathcal{F} - \{\varnothing\}$, and $A \subseteq B$. In terms of the earlier notation, in "$(a \mid a, b)$", a is focal event $\{a\}$ and a, b is the context event $\{a, b\}$, and thus $(a \mid a, b)$ is the same as $(\{a\} \mid \{a, b\})$. □

Definition 7.4 Throughout this chapter $\precsim$ denotes a binary relation of $\mathcal{F}$. Note that $\boldsymbol{\precsim}$ differs from $\precsim$ in that the former is in bold typeface. □

Also note that the Basic Belief and Basic Choice Axioms are concerned with finite conditional events of the form $(A \mid B)$, where $A = \{b\}$, $b \in B$, and B has at least 2 elements.

Axiom 7.1 *The following two statements hold:*

(1) $\boldsymbol{\precsim}$ *is a weak order on the set of finite conditional events of* X.

(2) $\boldsymbol{\precsim}$ *is an extension of* $\precsim$ *(where* $\precsim$ *is as in Axiom 6.1).* □

Axiom 7.2 *The following two statements hold for all finite conditional events* $(A \mid B)$ *and contexts* C *of* X*:*

(1) $(A \mid B) \sim (\varnothing \mid C)$ *iff* $A = \varnothing$.

(2) if $A \neq \varnothing$ *then* $(\varnothing \mid C) \prec (A \mid B)$. □

Axiom 7.3 *For each nonempty finite event* A *of* X *there exists* e *in* $X - A$ *such that*

(1) for each finite event B *of* X, *if* $\varnothing \subset A \subset B$ *(and therefore* B *has at least two elements), then there exists* E *in* $\mathcal{C}$ *such that* $B \sim_{\mathcal{C}} E$ *(Definition 6.3) and*

$$(A \mid B) \sim (e \mid E);$$

and

(2) there exists f *in* X *such that* $f \neq e$, $f \notin A$, *and*

$$(f \mid e, f) \sim (f \mid A \cup \{f\}). \quad \square$$

Axiom 7.4 *Suppose* $\varnothing \subset A \subset B$, $\varnothing \subset A \subset B'$, $e \in E$, $e' \in E'$, $B \sim_{\mathcal{C}} E$, $B' \sim_{\mathcal{C}} E'$, *and*

$$(A \mid B) \sim (e \mid E) \text{ and } (A \mid B') \sim (e' \mid E').$$

Then

$$(e \mid e, e') \sim (e' \mid e, e'). \quad \square$$

Axiom 7.3 associates with each finite conditional event $(A \mid B)$, where $A \neq \varnothing$ and B has at least two elements, a finite conditional event of the form $(e \mid E)$, where $e \in X$ and $E \in \mathcal{C}$, such that

$$(A \mid B) \sim (e \mid E).$$

Because $\boldsymbol{\precsim}$ is an extension of $\precsim$, this allows the placement of $(A \,|\, B)$ in the $\boldsymbol{\precsim}$-ordering to be determined by the placement of $(e \,|\, E)$ in the $\precsim$-ordering.

Axioms 7.3 and 7.4 play a critical role in extending a basic belief representation $\langle u, v \rangle$ for $\precsim$ to elements of the domain of $\boldsymbol{\precsim}$. Together, they allow each nonempty finite event A to be associated with an element e of X so that $u(A)$ can be defined as $u(e)$ and $v(A)$ as $v(e)$.

Axiom 7.5 *The following two statements are true:*

1. *For each a in X there exists E in $\mathcal{C}$ such that*

$$(\{a\} \,|\, \{a\}) \sim (E \,|\, E)\,.$$

2. *Let $(A \,|\, B)$ and $(C \,|\, D)$ be arbitrary finite conditional events such that B and D are in $\mathcal{C}$. Then there exist finite events B' and D' such that $B \cap B' = \varnothing$, $B \sim_{\mathcal{C}} B'$, $D \cap D' = \varnothing$, $D \sim_{\mathcal{C}} D'$, and*

$$(A \,|\, B) \boldsymbol{\precsim} (C \,|\, D) \text{ iff } (A \,|\, B \cup B') \boldsymbol{\precsim} (C \,|\, D \cup D')\,. \quad \Box$$

In Statement 2 of Axiom 7.5, $B \cup B'$ and $D \cup D'$ can be viewed intuitively as doubling respectively the "strengths" of B and D. Statement 2 then says that such doublings leave invariant comparisons of degrees of conditional belief. Statement 2 is obviously valid if degrees of conditional belief are measured by a Kolmogorov conditional probability function.

Definition 7.1 The *Belief Axioms* consist of the Basic Belief Axioms (Definition 7.1) together with Axioms 7.1 to 7.5. □

Definition 7.5 $\langle u, v \rangle$ is said to be a *belief representation for* $\boldsymbol{\precsim}$ if and only if the following four conditions hold:

(*i*) u and v are functions from $\mathcal{F}$ into respectively $\mathbb{R}^+ \cup \{0\}$ and $\mathbb{R}^+$.

(*ii*) For each $E \in \mathcal{F}$, $u(E) = 0$ if and only if $E = \varnothing$.

(*iii*) For all finite conditional events $(A \,|\, B)$,

$$u(B) = \sum_{b \in B} u(\{b\}) \text{ and } u(A) = \sum_{a \in A} u(\{a\}) \text{ if } A \neq \varnothing.$$

(*iv*) For all finite conditional events $(A \,|\, B)$ and $(C \,|\, D)$, and

$$(A \,|\, B) \boldsymbol{\precsim} (C \,|\, D) \text{ iff } \frac{u(A)}{u(B)} \cdot v(A) \leq \frac{u(C)}{u(D)} \cdot v(C)\,. \quad \Box$$

Condition (*iv*) of Definition 7.5 generalizes Definition 7.2 by having u be defined for focal events that are finite subsets of X. (In Definition 7.2, the focal events are singleton subsets of X.) Also note that u is additive for focal events, that is, if $A = \{a_1, \ldots, a_n\}$, then $u(A) = \sum_1^n u(a_i)$.

Theorem 7.3 *Assume the Belief Axioms (Definition 7.1). Then the following two statements hold:*

1. (Representation Theorem) *There exists a belief representation for* $\precsim$ *(Definition 7.5).*
2. (Uniqueness Theorem) *Let*

   $\mathcal{U} = \{u | \textit{there exists } v \textit{ such that } \langle u, v\rangle \textit{ is a belief representation for } \precsim\}$

   and

   $\mathcal{V} = \{v | \textit{there exists } u \textit{ such that } \langle u, v\rangle \textit{ is a belief representation for } \precsim\}$.

   Then $\mathcal{U}$ *and* $\mathcal{V}$ *are ratio scales.*

Proof. Theorem 3.1 of Narens (2003). □

Definition 7.6 Let $\langle u, v\rangle$ be a belief representation for $\precsim$. Then u is said to be the *context support function* of $\langle u, v\rangle$ and v the *definiteness function* of $\langle u, v\rangle$. □

Belief representations $\langle u, v\rangle$ for $\precsim$ with are useful in applications where non-probabilistic dimensions of uncertainty naturally arise. In this book, the non-probabilistic dimensions are called "definiteness" dimensions. Some of them may split into additional dimensions. In the intended interpretations, the context support function u measures the probabilistic dimension. The definiteness function, v, of course, measures definiteness.

In applications, one encounters various kinds of "definitenesses," and thus the intended interpretation of v may vary with the kind of definiteness encountered. One important kind that occurs in the decision literature is the ordinal opposite of "ambiguity" as described in the following quotation from Ellsberg (1961):

> Let us assume, for purposes of discussion, that an individual can always assign relative weights to alternative probability distributions reflecting the relative support given by his information, experience and intuition to these rival hypotheses. This implies that he can always assign relative likelihoods to

the states of nature. But how does he *act* in the presence of his uncertainty? The answer to that may depend on another sort of judgment, about the reliability, credibility, or adequacy of his information (including his relevant experience, advice and intuition) as a whole: not about the relative support it may give to one hypothesis as opposed to another, but about its ability to lend support to any hypothesis at all.

If all the information about the events in a set of gambles were in the form of sample-distributions, the ambiguity might be closely related, inversely to the size of the sample. But sample-size is not a universally useful index of this factor. Information about many events cannot be conveniently described in terms of a sample distribution; moreover, sample-size seems to focus mainly on the quantity of information. "Ambiguity" may be high (and the confidence in any particular estimate of probabilities low) even where there is ample quantity of information, when there questions of reliability and relevance of information, and particularly where there is *conflicting* opinion and evidence.

This judgment of the ambiguity of one's information, of the over-all credibility of one's composite estimates, of one's confidence in them, cannot be expressed in terms of relative likelihoods or events (if it could, it would simply affect the final, compound probabilities). Any scrap of evidence bearing on relative likelihood should already be represented in those estimates. But having exploited knowledge, guess, rumor, assumption, advice to arrive at a final judgment that one event is more likely than another or that they are equally likely, one can still stand back from this process and ask: "How much, in the end, is all this worth? How much do I really know about the problem? How firm a basis for choice, for appropriate decision and action, do I have?" The answer, "I don't know very much, and I can't rely on that," may sound familiar, even in connection with markedly unequal estimates of relative likelihood. If "complete ignorance" is rare or nonexistent, "considerable" ignorance is surely not. *(pp. 659–660)*

Definition 7.7 Suppose $\langle u, v\rangle$ is a belief representation for $\precsim$ with context function u and definiteness function v (Definition 7.5). Then $\mathbb{B}$ is said to be a *conditional belief function* for $\precsim$ if and only if for all finite conditional events $(A \mid B)$,

$$\mathbb{B}(A \mid B) = \frac{u(A)}{\sum_{b \in B} u(b)} v(A) . \quad \square \tag{7.8}$$

In Equation 7.8, u is often interpreted as a measure of probabilistic strength, and v as a measure of something that is the opposite of ambiguity or vagueness. Under these interpretations, the right-hand side of Equation 7.8 is interpreted as a subjective probability $\mathbb{P}(A \mid B)$ of A occurring when B is presented, where

$$\mathbb{P}(A \mid B) = \frac{u(A)}{\sum_{b \in B} u(b)},$$

weighted by the definiteness factor $v(A)$, that is,

$$\mathbb{B}(A \mid B) = \mathbb{P}(A \mid B) v(A) . \tag{7.9}$$

Definition 7.8 $\mathbb{P}$ in Equation 7.9 is called the $\mathbb{B}$-*subjective probability function* and v is called the $\mathbb{B}$-*definiteness function.* $\square$

Let $\mathcal{E}$ be a boolean algebra of events on Y. Let $\mathbf{P}$ be an arbitrary probability function on $\mathcal{E}$, and let $\mathbf{B}$ be any function on $\mathcal{E}$ into the nonnegative reals such that for all C and D in $\mathcal{E}$, if $C \subseteq D$ then $\mathbf{B}(C) \leq \mathbf{B}(D)$. For each nonempty A in $\mathcal{E}$, let

$$w(A) = \begin{cases} \frac{\mathbf{B}(A)}{\mathbf{P}(A)} & \text{if } \mathbf{P}(A) \neq 0 \\ 0 & \text{if } \mathbf{P}(A) = 0 . \end{cases}$$

Then $\mathbf{B}(A) = \mathbf{P}(A) w(A)$. Thus $\mathbf{B}$ having the form "$\mathbf{P}(A) w(A)$," puts almost no restriction on $\mathbf{B}$ given $\mathbf{P}$; that is, given a probability function $\mathbf{P}$ on $\mathcal{E}$, nearly any nonnegative real valued function $\mathbf{B}$ that is monotonic in terms of $\subseteq$ is obtainable through an appropriate choice of w. This does not mean that the representation $\mathbf{B}(A) = \mathbf{P}(A) w(A)$ is useless, because in many situations $\mathbf{P}$ is fixed and the form of w has to obey some qualitative restrictions.

However, conditional belief functions of the form,

$$\mathbb{B}(E \mid F) = \frac{\mathbb{P}(E) v(E)}{\mathbb{P}(F)} = \mathbb{P}(E \mid F) v(E),$$

where E and F are finite events of X such that $E \subseteq F$ and $F \neq \varnothing$, are greatly restricted. This is because $v(E)$ does not vary with the conditioning event F, which is a powerful restriction. In Section 7.5, section $v(E)$ is viewed as a factor that describes the amount the probability $\mathbb{P}(E \mid F)$ is distorted in a cognitive representation of $(E \mid F)$. Note that this way of viewing v requires that the distortion of $\mathbb{P}(E \mid F)$ in $\mathbb{B}(E \mid F)$ to be the same as the distortion of $\mathbb{P}(E \mid H)$ in $\mathbb{B}(E \mid H)$ for all H such that $E \subseteq H$.

Equation 7.9 allows for $\mathbb{B}(A \mid A) \neq 1$. In some situations this is desirable, while in others it is not. Nevertheless, even when Equation 7.9 provides an inadequate description of the degree of belief for many events of the form $(C \mid C)$, it can still be useful for the analysis and modeling of other kinds of more interesting conditional events.

For example, psychological studies of probability judgments focuses on uncertainty instead of certainty and therefore almost never present conditional events of the form $(A \mid A)$ to participants. Present day psychologists consider judgments of probability result from applying heuristics to various probabilistic situations. The kind of heuristic employed often varies with situation. The empirical evidence presented in this chapter and Chapter 10 suggest that there are considerable structural regularities in human probability judgments for events displaying a reasonable amount of uncertainty that run counter to the Kolmogorov theory. In such situations, Equation 7.9 is often applicable for accounting for the structural regularities. These regularities do not extend to certain or near certain events, because the participants apply different heuristics to such events, for example, they immediately see from the event's description that it is certain and assign it degree of belief 1, avoiding the kind of the heuristics they apply to reasonably uncertain events.[1]

Also, it is worthwhile noting that many kinds of probabilistic situations studied in the belief literature involve analogs of unconditional probability; that is, situations with a sure event and only evaluations are made about events conditioned on the sure event. In such situations, if the sure event has definiteness 1, then evaluations of the form $\mathbb{B}(A \mid A) \neq 1$ for non-sure events never occur.

7.3 Choice Axioms

Definition 7.9 The *Choice Axioms consist of the Belief Axioms with the*

[1]Equation 7.9 is derived from the Belief Axioms which assume the existence of events of the form $(A \mid A)$. However, Equation 7.9 can also be derived from a reformulation of the Belief Axioms that eliminates such events. That is, instead of extending the Basic Belief Axioms to all finite conditional events of X (as is done in the Belief Axioms), they can be extended to the subset of finite conditional events that correspond to the kinds of "conditional events with reasonable uncertainty" generally used in psychological experimentation. This is not difficult to accomplish, and using the method of proof of Theorem 7.3, a result analogous to Theorem 7.3 easily follows.

axiom of Binary Symmetry. □

Theorem 7.4 *Assume the Choice Axioms. Let* $\mathbb{B}$ *be a belief representation for* $\precsim$ *with* $\mathbb{B}$*-subjective probability function* $\mathbb{P}$ *and* $\mathbb{B}$*-definiteness function* v*. Then for all finite conditional events* $(A \mid B)$ *and* $(C \mid D)$ *of* X*,*

$$(A \mid B) \precsim (C \mid D) \text{ iff } \mathbb{P}(A \mid B) \leq \mathbb{P}(C \mid D).$$

Proof. Theorem 4.1 of Narens (2003). □

7.4 Belief Support Functions

Convention 7.2 Suppose a system $\mathcal{S}$ of conditional probabilities satisfies Luce's Choice Axiom. Then by Theorem 6.1, there exists a ratio scale family of functions $\mathcal{U}$ such that for each u in $\mathcal{U}$ and each $\mathbb{P}(A|B)$ in $\mathcal{S}$,

$$\mathbb{P}(A \mid B) = \frac{\sum_{a \in A} u(a)}{\sum_{b \in B} u(b)}.$$

In the above choice situation, we start with $\mathcal{S}$ and construct $\mathcal{U}$. But the reverse procedure is also important, where we start with a ratio scale family of functions $\mathcal{U}$, and then construct the system $\mathcal{S}$ of conditional probabilities satisfying Luce's Choice Axiom by,

$$\mathbb{P}(C \mid D) = \frac{\sum_{c \in C} u(c)}{\sum_{d \in D} u(d)},$$

where C and D are finite events and $u \in \mathcal{U}$. In the reverse procedure, $\mathcal{U}$ is called a *support family* and elements of $\mathcal{U}$ are called *support functions.* The intended interpretation of $\mathcal{U}$ is that it is a family of functions that measure the support for the occurrences of events based on evidence about the events; that is, for the finite event C, $u(C) = \sum_{c \in C} u(C)$ is a measurement by $u \in U$ of how well the evidence supports the occurrence of C. □

Let $\langle u, v \rangle$ be a belief representation for $\precsim$. Instead of using u to measure the support for finite events, the function $\mathfrak{s}(x) = u(x)v(x)$, $x \in X$, is used for measuring support. Thus $\mathfrak{s}$ incorporates measurements from both the the probabilistic and definiteness dimensions.

Definition 7.10 $\mathfrak{s}$ is said to be a *belief support function* for $\precsim$ if and only if there exists a belief representation $\langle u, v \rangle$ for $\precsim$ such that for all finite subsets A of X,

$$\mathfrak{s}(A) = u(A)v(A).$$

It easily follows from Theorem 7.3 that if the Belief Axioms hold, then the set of all belief support functions for $\precsim$ is a ratio scale. □

Definition 7.11 Let $\mathfrak{s}$ be the belief support function for $\precsim$. Then $\mathbb{P}_{\mathfrak{s}}$ is said to be the *subjective support probability function* for $\precsim$ if and only if $\mathbb{P}_{\mathfrak{s}}$ is the function on the finite conditional events of X such that for each finite conditional event $(A \mid B)$ of X,

$$\mathbb{P}_{\mathfrak{s}}(A \mid B) = \frac{\mathfrak{s}(A)}{\mathfrak{s}(A) + \mathfrak{s}(B - A)}.$$

Note that if the Belief Axioms hold, then there is only one subjective support probability function for $\precsim$, that is, if $\mathfrak{s}$ and $\mathfrak{t}$ are belief support functions for $\precsim$, then $\mathbb{P}_{\mathfrak{s}} = \mathbb{P}_{\mathfrak{t}}$. $\square$

Also note that $\mathbb{P}_{\mathfrak{s}}$ has the following property of probability functions: For disjoint finite events A and B,

$$\mathbb{P}_{\mathfrak{s}}(A \mid A \cup B) + \mathbb{P}_{\mathfrak{s}}(B \mid A \cup B) = 1.$$

However, unlike a probability function, $\mathbb{P}_{\mathfrak{s}}$ may not be *additive;* that is, there are situations with finite events C, D, and E such that $C \cap D = \varnothing$ and

$$\mathbb{P}_{\mathfrak{s}}(C \cup D \mid E) \neq \mathbb{P}_{\mathfrak{s}}(C \mid E) + \mathbb{P}_{\mathfrak{s}}(D \mid E).$$

Often in making a generalization some concepts split into two or more concepts. This is the case with generalizing choice "representation" to "belief representation." The concept of conditional probability of A given B, $\mathbb{P}(A \mid B)$, in the choice representation naturally splits in the belief representation into the concepts of conditional belief of A given B, $\mathbb{B}(A \mid B)$, and belief support conditional probability of A given B, $\mathbb{P}_{\mathfrak{s}}(A \mid B)$. When the definiteness function v is constant on finite events, $\mathbb{B}$ and $\mathbb{P}_{\mathfrak{s}}$ agree and degenerate to a multiple of $\mathbb{P}$, or equivalently, when v is constant on finite events, $\mathbb{B}$ is a multiple of $\mathbb{P}_{\mathfrak{s}}$.

$\mathbb{B}$ and $\mathbb{P}_{\mathfrak{s}}$ are related by the following: For all finite conditional events $(A \mid B)$ of X,

$$\mathbb{P}_{\mathfrak{s}}(A \mid B) = \frac{\mathbb{B}(A \mid B)}{\mathbb{B}(A \mid B) + \mathbb{B}((B - A) \mid B)}.$$

For a task where an individual is asked to judge numerically the probabilities of conditional events, both $\mathbb{B}$ and $\mathbb{P}_{\mathfrak{s}}$ are natural candidates for modeling the judgment. For the empirical results discussed in Section 7.5, $\mathbb{P}_{\mathfrak{s}}$ is better in this regard.

7.4.1 Fair Bets

The notion of a "fair bet" relates the strengths of beliefs of the events in the bet to the value of the outcomes of the event. The following is one reasonable notion of "fair bet":

Definition 7.12 Let $\mathbb{B}$ be an individual's conditional belief function for $\precsim$, $\mathbb{P}$ his or her $\mathbb{B}$-subjective probability function, and A and B be nonempty finite events such that $A \cap B = \varnothing$. Consider the following two gambles:

H_1: The individual gains something that has value $a > 0$ if A occurs, and loses something that has value $b > 0$ if B occurs.

H_2: The individual loses something that has value $a > 0$ if A occurs, and gains something that has value $b > 0$ if B occurs.

Then H_1 and H_2 are said to be a *fair bet pair* if and only if

$$a\mathbb{P}_{\mathfrak{s}}(A \mid A \cup B) - b\mathbb{P}_{\mathfrak{s}}(B \mid A \cup B) = 0. \tag{7.10}$$

H is said to be a *fair bet* if and only if there exists an H' such that H and H' are a fair bet pair. □

The idea of the notion of "fair bet" in Definition 7.12 is that the individual, if forced to bet, is indifferent to which side of the bet he takes. In Definition 7.12, H_1 and H_2 express the two sides of the bet. If in Equation 7.10 a were increased a little to a^+, then he or she would prefer betting on A, that is, prefer the gamble H_1; if b is increased a little to b^+ in Equation 7.10, then he or she would prefer the gamble H_2. Note that "fair bet" as defined in Definition 7.12 is equivalent to the following formulation: H_1 and H_2 are a fair bet pair if and only if

$$\frac{\mathbb{B}(A \mid A \cup B)}{\mathbb{B}(B \mid A \cup B)} = \frac{b}{a}.$$

Let $\mathbb{B}$ be an individual's conditional belief function for $\precsim$, $\mathbb{P}$ his $\mathbb{B}$-subjective probability function, u his $\mathbb{B}$-probabilistic support function, and v his $\mathbb{B}$-definiteness function. Let A, B, C, and D be nonempty finite events such that

$$(A \mid A \cup B) \sim (B \mid A \cup B) \text{ and } (C \mid C \cup D) \sim (D \mid C \cup D). \tag{7.11}$$

Assume that a high definiteness value, say 1, is assigned to A and B because much is known about them and a lower definiteness value, $\alpha < 1$, is assigned to C and D, because little is known about them. Then by Equation 7.11,

$$u(A) = u(B) \text{ and } u(C) = u(D).$$

Then,

$$\mathbb{B}(A \mid A \cup B) = \mathbb{B}(B \mid A \cup B) = \frac{1}{2} \text{ and } \mathbb{B}(C \mid C \cup D) = \mathbb{B}(D \mid C \cup D) = \frac{1}{2}\alpha\,.$$

Therefore,

$$\mathbb{P}(A \mid A \cup B) = \mathbb{P}(C \mid C \cup D) = \frac{1}{2}$$

and

$$\mathbb{P}_{\mathfrak{s}}(A \mid A \cup B) = \mathbb{P}_{\mathfrak{s}}(C \mid C \cup D) = \frac{1}{2}.$$

Thus, although the conditional events $(A \mid B)$ and $(C \mid D)$ differ in degree of belief and definiteness, they are given the same value by the subjective support probability function $\mathbb{P}_{\mathfrak{s}}$ and the $\mathbb{B}$-subjective probability function $\mathbb{P}$.

In general for conditional events $(E \mid F)$ with $v(E) = v(F)$,

$$\mathbb{P}(E \mid F) = \mathbb{P}_{\mathfrak{s}}(E \mid F)\,.$$

Thus, to differentiate $\mathbb{P}_{\mathfrak{s}}$ from $\mathbb{P}$ (and therefore, from $\mathbb{B}$) in an interesting manner, one needs to consider a situation where the definitenesses of the focal event differs from the definitenesses of its context event.

7.4.2 Ellsberg's Paradox

A famous example of such a situation is supplied by Ellsberg (1961). Suppose an urn has 90 balls that have been thoroughly mixed. Each ball is of one of the three colors, red, blue, or yellow. There are thirty red balls, but the number of blue balls and the number of yellow balls are unknown except that together they total 60. A ball is to be randomly chosen from the urn. Let R be the event that a red ball is chosen, B the event a blue ball is chosen, and Y the event a yellow ball is chosen. Let $U = R \cup B \cup Y$. Assume that this situation is part of the domain of a conditional belief function $\mathbb{B}$. Let $\mathbb{P}$ be the $\mathbb{B}$-subjective probability function and v the $\mathbb{B}$-definiteness function. The following are reasonable assignments of $\mathbb{B}$-probabilities and $\mathbb{B}$-definitenesses:

$$\mathbb{P}(R \mid U) = \mathbb{P}(B \mid U) = \mathbb{P}(Y \mid U) = \frac{1}{3}\,,$$

$$\mathbb{P}(R \cup Y \mid U) = \mathbb{P}(B \cup Y \mid U) = \mathbb{P}(R \cup B \mid U) = \frac{2}{3}\,,$$

and

$$v(B) = v(Y) < v(R \cup Y) = v(R \cup B) < v(B \cup Y) = v(R)\,.$$

It is easy to verify that the following three statements hold:

1. $\mathbb{B} \neq \mathbb{P}$,
2. $\mathbb{B}(B \,|\, U) < \mathbb{B}(R \,|\, U)$ and $\mathbb{B}(B \cup Y \,|\, U) > \mathbb{B}(R \cup Y \,|\, U)$.
3. $\mathbb{P}_{\mathfrak{S}}(B \,|\, U) < \mathbb{P}_{\mathfrak{S}}(R \,|\, U)$ and $\mathbb{P}_{\mathfrak{S}}(B \cup Y \,|\, U) < \mathbb{P}_{\mathfrak{S}}(R \cup Y \,|\, U)$.

Ellsberg (1961) and many subsequent experiments have verified that most subjects prefer the gamble G_2 to G_1, where

$$G_1 = \begin{cases} \text{Receive \$100 if } B \\ \text{Receive \$0 if } R \cup Y \end{cases} \qquad G_2 = \begin{cases} \text{Receive \$100 if } R \\ \text{Receive \$0 if } B \cup Y\,. \end{cases}$$

This is consistent with the preference being determined by the degree of belief given in the first part of Statement 2. Most of these subjects (as well as most subjects in general) prefer gamble H_2 to H_1, where

$$H_1 = \begin{cases} \text{Receive \$100 if } R \cup Y \\ \text{Receive \$0 if } B \end{cases} \qquad H_2 = \begin{cases} \text{Receive \$100 if } B \cup Y \\ \text{Receive \$0 if } R\,, \end{cases}$$

which is consistent with the preference being determined by the degree of belief given in the second part of Statement 2. Ellsberg and many other researchers believe that the above preferences do not violate rationality, that is, it is proper to incorporate "ambiguity" as factor in the evaluation of the value of a gamble. Other researchers disagree.

It should be noted that because each of the above gambles has outcomes \$100 and \$0, the preferences over the two pairs of gambles can be decided by directly appealing to the ordering of degrees of belief of the appropriate events. Thus in particular, these preferences can be determined without computing the values of each gamble and then determining preference by ordering the gambles, that is, the formula for determining the value of a gamble in terms of $\mathbb{B}$ and v and the values of \$100 and \$0 need not be known to appropriately order the gambles for risk adverse individuals.

Ellsberg's Paradox is about unconditional probability rather than conditional probability, because all the judgments concern a degenerate case of conditional probability where all the conditioning events are identical. Also, the number of balls in the conditioning event U is nonambiguously 90. It is interesting to theorize about the impact of extending Ellsberg's concept of ambiguity to conditional probability including cases where the "size" of the conditioning event is ambiguous.

Consider the following example: There are black balls, gray balls, orange balls, and a white ball. There are 8 orange balls, 1 white ball, but the number of black and gray balls are unknown except that their total is 2000. Three cases of conditional probabilities of a ball drawn from an urn are considered. Let B, W, G, and O stand for, respectively, the events of a black, white, gray, and orange ball being chosen.

Case 1. The black and gray balls are put into an empty urn. Then applying the conditional belief function $\mathbb{B}$ with $\mathbb{B}$-definiteness, say, $v(B) = .8$ and $\mathbb{B}$-probability, say, $\mathbb{P}(B \mid B \cup G) = .5$ yields $\mathbb{B}(B \mid B \cup G) = .4$.

Case 2. The white ball and all the black balls are put into an empty urn. Then using the $\mathbb{B}$-definiteness $v(B) = .8$ from Case 1 for the event B and setting $v(W) = 1$ then yields,

$$\mathbb{B}(B \mid B \cup W) = \mathbb{P}(B \mid B \cup W)v(B) = \mathbb{P}(B \mid B \cup W) \cdot .8 \leq .8\,,$$

because $\mathbb{P}(B \mid B \cup W) \leq 1$.

Case 3. The white ball and the 8 orange balls are put into an empty urn. Then setting $v(O) = 1$ yields,

$$\mathbb{B}(O \mid O \cup W) = \mathbb{P}(O \mid O \cup W) \cdot v(O) = \frac{8}{9} \cdot 1 = \frac{8}{9}\,.$$

However, intuitively it should be the case that

$$\mathbb{B}(B \mid B \cup W) > \mathbb{B}(O \mid O \cup W)\,.$$

(If the reader's intuition is different, then the reader should try to play with the numbers used in the example, e.g., let the total number of black and gray balls be 2,000,000, until an example can be found that works for the reader.) But this intuitive result contradicts results from Cases 2 and 3 that imply

$$\mathbb{B}(B \mid B \cup W) < \mathbb{B}(O \mid O \cup W)\,.$$

These considerations indicate to me that an incorrect model is being used for the situation described in the last example. The incorrectness is due, in my opinion, to having the definiteness of B not vary with the conditioning event. Thus for the form of Ellsberg's ambiguity considered in the last example, the belief function $\mathbb{B}$ needs to be generalized, for example, generalized in a manner so that the definiteness function depends in part on the conditioning event. There are many ways to accomplish this, and it is not clear at this time which approaches to generalization are best.

Thus in summary, the conditional belief representation accounts for the Ellsberg Paradox. However, if the Paradox is extended to situations where the size of the conditioning event is ambiguous, then, from a theoretical perspective, it is unlikely that the conditional belief representation will account for the extended version. Qualitatively, this means that for extended version, not only binary symmetry—which is characteristic of the kind of ambiguity inherent in the Ellsberg Paradox—fails but one of the Belief Axioms must also fail.

7.5 Application to Support Theory

Belief support probability functions provide a theory of subjective probability judgments that bears many similarities with the descriptive theory of human probability judgments called *Support Theory* developed in Tversky and Koehler (1994) and Rottenstreich and Tversky (1997). Support theory was designed by Tversky and Koehler (1994) to explain various phenomena observed in empirical studies of human judgments of probability. These studies usually showed that the sum of the probabilities of a binary partition of an event was near 1, but when one of the partitions was further partitioned, then the sum of the probabilities over the expanded partition was significantly > 1. The following study of Fox and Birke (2002) is a typical example of this phenomenon.

7.5.1 Empirical Phenomena

Example 7.1 (Jones vs. Clinton) 200 practicing attorneys were recruited (median reported experience: 17 years) at a national meeting of the American Bar Association (in November 1997). 98% of them reported that they knew at least "a little" about the sexual harassment allegation made by Paula Jones against President Clinton. At the time that the survey, the case could have been disposed of by either

(A) judicial verdict or

(B) an outcome other than a judicial verdict.

Furthermore, outcomes other than a judicial verdict (B) included

(B1) settlement;

(B2) dismissal as a result of judicial action;

(B3) legislative grant of immunity to Clinton; and

(B4) withdrawal of the claim by Jones. □

Each attorney was randomly assigned to judge the probability of one of these six events. The results are given in Table 7.1

(A) judicial verdict	.20
(B) not verdict	.75
Binary partition total	.95
(A) judicial verdict	.20
(B1) settlement	.85
(B2) dismissal	.25
(B3) immunity	.0
(B4) withdrawal	.19
Five fold partition total	1.49

Table 7.1: Median Judged Probabilities for All Events in Study

7.5.2 Tversky's and Koehler's Theory

Support theory has its own approach and terminology for modeling probability judgments.

Convention 7.3 Participants' probability judgments are made on descriptions of events, called *hypotheses*, instead of events. A finite set T with at least two elements is assumed. This set generates an event space. A set of hypotheses $\mathcal{H}$ is assumed such that each hypothesis A in $\mathcal{H}$ describes an event, called the *extension of* A and denoted by A', that is a subset of T. It is allowed that different hypotheses describe the same event; for example, for a roll of a pair of dice, the hypotheses "the sum is 3" and "the product is 2" describe the same event, namely one die shows 1 and the other 2.

Hypotheses that describe an event $\{t\}$, where $t \in T$, are called *elementary*; those that describe $\varnothing$ are called *null*; and nonnull hypotheses A and B in $\mathcal{H}$ such that the conjunction of A and B describe $\varnothing$ are called *exclusive (with respect to* $\mathcal{H}$*)*. C in $\mathcal{H}$ is said to be an *explicit disjunction (with respect to* $\mathcal{H}$*)*—or for short, an *explicit hypotheses (of* $\mathcal{H}$*)*—if and only if there are exclusive A and B in $\mathcal{H}$ such that $C = A \vee B$, where "$\vee$" stands for the logical disjunction of A and B, "A or B." D in $\mathcal{H}$ is said to be *implicit (with respect to* $\mathcal{H}$*)* if and only if D is not $\varnothing$, is not elementary, and is not explicit with respect to $\mathcal{H}$.

$\mathcal{H}$ may have implicit and explicit hypotheses that describe the same event. For example, consider the following descriptions:

C: "Ann majors in a natural science."

A: "Ann majors in a biological science."

B: "Ann majors in a physical science."

Then C and $A \vee B$ describe the same event, that is, have the same *extension*, or using the convention of H' standing for the extension of the hypothesis H,

$$C' = (A \vee B)' = A' \cup B'.$$

It is assumed that whenever exclusive A and B belong to $\mathcal{H}$, then their disjunction $A \vee B$ also belong to $\mathcal{H}$. $\square$

Tversky and Koehler (1994) provided empirical data from many experiments where participants judged explicit hypotheses E to be more likely than implicit ones I with the same extensions ($E' = I'$). They suggested that this empirical result reflected a basic principle of human judgment. They explained the result in terms of an intuitive theory of information processing that involved (i) the formation of a "global impression that is based primarily on the most representative or available cases" and modulated by factors such as memory and attention, and (ii) the making of judgments that are mediated by heuristics such as representativeness, availability, and anchoring and adjusting. (The heuristics in (ii) are described in Subsection 10.2.1.)

Formally, support theory is formulated in terms of "evaluation frames" and "support functions:"

> An *evaluation frame* (A, B) consists of a pair of exclusive hypotheses: the first element A is the focal hypothesis that the judge evaluates, and the second element B is the alternative hypothesis. We assume that when A and B are exclusive the judge perceives them as such, but we do not assume that the judge can list all the constituents of an implicit disjunction. Thus, the judge recognizes the fact that "biological sciences" and "physical sciences" are disjoint categories, but he or she may be unable to list all their disciplines. This is a form of bounded rationality; we assume recognition of exclusivity, but not perfect recall.
>
> We interpret a person's probability judgment as a mapping P from an evaluation frame to the unit interval. ... Thus, $P(A, B)$ is the judged probability that A rather than B holds, assuming that one and only one of them is valid. ...

An *evaluation frame* (A, B) consists of a pair of exclusive hypotheses: The first element A is the *focal* hypothesis that the judge evaluates, and the second element B is the *alternative* hypothesis. To simplify matters, we assume that when A and B are exclusive, the judge perceives them as such, but we do not assume that the judge can list all the constituents of an implicit disjunction. In terms of [the above example involving Ann], we assume that the judge knows, for instance, that genetics is a biological science, that astronomy is a physical science, and that the biological and the physical sciences are exclusive. Thus, we assume recognition of inclusion but not perfect recall.

We interpret a person's probability judgment as a mapping P from an evaluation frame to the unit interval. To simplify matters we assume that $P(A, B)$ equals zero if and only if A is null and that it equals one if and only if B is null; we assume that A and B are both not null. Thus, $P(A, B)$ is the judged probability that A rather than B holds, assuming that one and only one of them is valid. Obviously, A and B may each represent an explicit or an implicit disjunction. The extensional counterpart of $P(A, B)$ in the standard theory is the conditional probability $P(A' \mid A' \cup B')$. The present treatment is nonextensional because it assumes that probability judgment depends on the descriptions A and B, not just on the events A' and B'. We wish to emphasize that the present theory applies to the hypotheses entertained by the judge, which do not always coincide with the given verbal descriptions. A judge presented with an implicit disjunction may, nevertheless, think about it as an explicit disjunction, and vice versa.

Support theory assumes that there is a ratio scale s (interpreted as degree of support) that assigns to each hypothesis in $\mathcal{H}$ a nonnegative real number such that for any pair of exclusive hypotheses A and B in $\mathcal{H}$,

$$P(A, B) = \frac{s(A)}{s(A) + s(B)}, \tag{7.12}$$

[and] for all hypotheses A, B, and C in $\mathcal{H}$, if B and C are exclusive, A is implicit, and $A' = (B \vee C)'$, then

$$s(A) \leq s(B \vee C) = s(B) + s(C). \tag{7.13}$$

Equation [7.12] provides a representation of subjective probability in terms of the support of the focal and the alternative hypotheses. Equation [7.13] states that the support of an implicit disjunction A is less than or equal to that of a coextensional explicit disjunction $B \vee C$ that equals the sum of the support of its components. Thus, support is additive for explicit disjunctions and subadditive for implicit ones. *(Tversky and Koehler, 1994, pp. 548–549.)*

Tversky and Koehler show that Equations 7.12 and 7.13 above imply the following four principles for all A, B, C, and D in $\mathcal{H}$:

1. *Binary Complementarity.* $P(A, B) + P(B, A) = 1$.

2. *Proportionality.* If A, B, and C are mutually exclusive and B is not null, then
$$\frac{P(A,B)}{P(B,A)} = \frac{P(A, B \vee C)}{P(B, A \vee C)}.$$

3. *Product Rule.* Let $R(A, B)$ be the odds of A against B, that is, let
$$R(A,B) = \frac{P(A,B)}{P(B,A)}.$$
Then
$$R(A,B)R(C,D) = R(A,D)R(C,B), \tag{7.14}$$
provided A, B, C, D are not null, and the four pairs of hypotheses in Equation 7.14 are pairwise exclusive.

4. *Unpacking Principle.* Suppose B, C, and D are mutually exclusive, A is implicit, and $A' = (B \vee C)'$. Then
$$P(A, D) \leq P(B \vee C, D) = P(B, C \vee D) + P(C, B \vee D).$$

Tversky and Koehler show the following theorem.

Theorem 7.5 *Suppose $P(A, B)$ is defined for all exclusive A and B in $\mathcal{H}$, and that it vanishes if and only if A is null. Then Binary Complementarity, Proportionality, the Product Rule, and the Unpacking Principle hold if and only if there exists a ratio scale s on $\mathcal{H}$ that satisfies Equations 7.12 and 7.13.*

Proof. Theorem 1 of Tversky and Koehler (1994). □

Narens (2003) used belief support probabilities to account for phenomena that form the basis of Tversky's and Koehler's support theory. This was accomplished by interpreting definiteness as a measure of unpackedness. In order to carry this out in terms of the ideas presented in this chapter, a few minor modifications to Tversky's and Koehler's setup are needed. (The purpose of the modifications is to make the discussion throughout the section coordinate to the discussions and results given in the previous sections of the chapter. They are not essential for the points made throughout the section.)

Convention 7.4 Instead of the finite set T of elementary hypotheses, an infinite set X of elementary hypotheses is assumed. It is also assume that $\mathcal{H}$ is a set of hypotheses that is (i) closed under disjunction, and (ii) is such that each hypothesis A in $\mathcal{H}$ has an extension A' that is a finite subset of X. To simplify notation and exposition, it is assumed that elements of $\mathcal{H}$ are nonnull. Instead of the evaluation frame notation (A, B), the conditional hypothesis notation $(A|A \vee B)$ (where, of course, $A \vee B$ is explicit) is employed to describe the kinds of situations encompassed by support theory. □

Convention 7.5 Throughout the remainder of this section the following notation and conventions are assumed.

- u is a function from X into the positive reals. u is extended to $\mathcal{H}$ as follows: For each nonempty finite subset A of X, let

$$u(A) = \sum_{a \in A} u(a) . \tag{7.15}$$

- For each H in $\mathcal{H}$, let $u(H) = u(H')$ (where, of course, H' is the extension of H). u is to be interpreted as a measure of probabilistic strength.

- v is a function from $\mathcal{H}$ into the positive reals such that for all A and B in $\mathcal{H}$, if A is implicit, B is explicit, and $A' = B'$, then

$$v(A) \leq v(B) . \tag{7.16}$$

  v is to be interpreted as a "distortion factor" due to specific kinds of cognitive processing, and Equation 7.16 captures the important characteristic of the distortion that is due to "unpacking."

- $\mathbb{B}$ is a function from conditional hypotheses to the positive reals such that for each conditional hypothesis $(A|A \vee B)$,

$$\mathbb{B}(A|A \vee B) = \frac{u(A)}{u(A \vee B)} \cdot v(A) \, . \quad \square \tag{7.17}$$

Equations 7.15 and 7.17 provide $\mathbb{B}$ with the same algebraic form as the conditional belief functions. The intended interpretation of $\mathbb{B}(A|A \vee B)$ is a distortion (by a factor of $v(A)$) of the probabilistic strength,

$$\frac{u(A)}{u(A \vee B)},$$

of $(A|A \vee B)$. In this section, the distortion is due to the Unpacking Principle, which is captured in large part by Equation 7.16.

Convention 7.6 Throughout the rest of the section, $\mathbb{P}_{\mathfrak{s}}$ is the $\mathbb{B}$-probabilistic support probability function; that is,

$$\mathbb{P}_{\mathfrak{s}}(A|A \vee B) = \frac{u(A)v(A)}{u(A)v(A) + u(B)v(B)} = \frac{\mathbb{B}(A|A \vee B)}{\mathbb{B}(A|A \vee B) + \mathbb{B}(B|A \vee B)} \, . \quad \square$$

It is easy to verify that $\mathbb{P}_{\mathfrak{s}}$ satisfies Binary Complementarity, Proportionality, and the Product Rule. The following theorem is immediate:

Theorem 7.6 *Suppose A, B, C, and D are arbitrary elements of $\mathcal{H}$, A and D are exclusive, B, C, and D are mutually exclusive, A is implicit, and $A' = (B \vee C)'$. Then the following three statements are true:*

1. *(*Ordinal Unpacking*)* $\mathbb{P}_{\mathfrak{s}}(A|A \vee D) \leq \mathbb{P}_{\mathfrak{s}}(B \vee C|B \vee C \vee D)$.
2. *(*Definiteness Unpacking*) If $v(B) = v(C)$, then*

$$\begin{aligned} \mathbb{P}_{\mathfrak{s}}(A|A \vee D) &\leq \mathbb{P}_{\mathfrak{s}}(B \vee C|B \vee C \vee D) \\ &= \mathbb{P}_{\mathfrak{s}}(B|B \vee C \vee D) + \mathbb{P}_{\mathfrak{s}}(C|C \vee B \vee D) \, . \end{aligned}$$

3. *The following two statements are equivalent:*

   *(*i*)* $\mathbb{B}$ *is* additive, *that is,*

$$\mathbb{B}(B \vee C|B \vee C \vee D) = \mathbb{B}(B|B \vee C \vee D) + \mathbb{B}(C|B \vee C \vee D) \, .$$

   *(*ii*) The Unpacking Principle holds, that is,*

$$\begin{aligned} \mathbb{P}_{\mathfrak{s}}(A|A \vee D) &\leq \mathbb{P}_{\mathfrak{s}}(B \vee C|B \vee C \vee D) \\ &= \mathbb{P}_{\mathfrak{s}}(B|B \vee C \vee D) + \mathbb{P}_{\mathfrak{s}}(C|C \vee B \vee D) \, . \quad \square \end{aligned}$$

Observe that the inequalities in Theorem 7.6 become strict if the inequality in Equation 7.16 becomes strict.

Narens (2003) comments the following about Theorem 7.6:

> The above shows that $\mathbb{B}$-probabilistic support probabilities, when generated by an additive $\mathbb{B}$, is a form of support theory. This form of support theory generalizes phenomena outside of support theory when generated by a non-additive $\mathbb{B}$. The non-additive case still retains much of the flavor of support theory, particularly its empirical basis, because Ordinal Unpacking and Definiteness Unpacking hold. Of possible empirical importance is the consideration that it may be possible to find many natural situations in which Definiteness Unpacking holds but the Unpacking Principle fails. *(pp. 13–14)*

Let $(A|A \vee B)$ be a conditional hypothesis. Then in the formula,

$$\mathbb{B}(A|A \vee B) = v(A)\frac{u(A)}{u(A \vee B)},$$

may be viewed as a "distortion" of the fraction,

$$\frac{u(A)}{u(A \vee B)}.$$

By the definition of u,

$$\frac{u(A)}{u(A \vee B)} = \frac{u(A')}{u((A \vee B)')} = \frac{u(A')}{u(A' \cup B')}.$$

However, by Equation 7.15,

$$\frac{u(A')}{u(A' \cup B')} = \frac{\sum_{a \in A'} u(a)}{\sum_{c \in A' \cup B'} u(c)},$$

may be interpreted as a subjective conditional probability of the conditional, extensional hypothesis $(A'|A' \cup B')$. Under this interpretation, $v(A)$ is the amount that the subjective conditional probability of the extension of $(A|A \vee B)$ needs to be distorted to achieve $\mathbb{B}(A|A \vee B)$.

Tversky and Koehler (1994) provided many examples of their theory (some of which are presented in Chapter 10). However, empirical results of Rottenstreich and Tversky showed that certain aspects of the Tversky and Koehler theory needed generalization.

7.5.3 Rottenstreich's and Tversky's Theory

Rottenstreich and Tversky (1997) constructed experiments in which the probabilities for explicit disjunctions $G \vee H$ were subadditive, that is, they found situations where sometimes

$$P(G \vee H) < P(G) + P(H).$$

They generalized Tversky and Koehler (1994) to accommodate subadditivity as follows: For all hypotheses E and F, where F is nonnull, let

$$R(E, F) = \frac{P(E, F)}{P(F, E)}.$$

Then for all hypotheses A, A_1, A_2, B, C, and D, the following three assumptions hold :

1. *(Binary Complementarity)* $P(A, B) + P(B, A) = 1$.
2. *(Product Rule)* (i) If (A, B), (B, D), (A, C), and (C, D) are exclusive, then
$$R(A, B)R(B, D) = R(A, C)R(C, D),$$
and (ii) if (A, B), (B, D), and (A, D) are exclusive, then
$$R(A, B)R(B, D) = R(A, D).$$
3. *(Odds Inequality)* Suppose A_1, A_2, and B are mutually exclusive, A is implicit, and the judge recognizes $A_1 \vee A_2$ as a partition of A. That is, $(A_1 \vee A_2)' = A'$ and the judge recognizes that $A_1 \vee A_2$ has the same extension as A. Then
$$R(A, B) \leq R(A_1 \vee A_2, B) \leq R(A_1, B) + R(A_2, B).$$

Rottenstreich and Tversky (1997) show the following:

Theorem 7.7 *Suppose $P(A, B)$ is defined for all exclusive hypotheses A and B and that it vanishes if and only if A is null. Then the above three assumptions hold if and only if there exists a nonnegative function s on the set of hypotheses such that for all exclusive hypotheses C and D,*

$$P(C, D) = \frac{s(C)}{s(C) + s(D)}.$$

Furthermore, if A_1 and A_2 are exclusive, A is implicit, and $(A_1 \vee A_2)$ is recognized as a partition of A, then

$$s(A) \leq s(A_1 \vee A_2) \leq s(A_1) + s(A_2). \quad \square$$

In Rottenstreich's and Tversky's theory, Binary Complementarity is the same as in Tversky and Koehler (1994). Their Product Rule is slightly stronger than in Tversky and Koehler, because it has the additional product condition $R(A, B)R(B, D) = R(A, D)$. Odds Inequality replaces the unpacking condition of Tversky and Koehler. Tversky and Rottenstreich (1997) note that in Odds Inequality, "The recognition requirement, which restricts the assumption of implicit subadditivity, was not explicitly stated in the original [Tversky and Koehler, 1994] version of the theory, although it was assumed in its applications."

Theorem 7.6 showed that $\mathbb{B}$-probabilistic support probabilities for additive $\mathbb{B}$ was a version of Tversky's and Koehler's support theory. However, for subadditive $\mathbb{B}$, the $\mathbb{B}$-probabilistic support probabilities naturally generalize to produce a theory similar to Rottenstreich's and Tversky's support theory, with the Product Rule and Odds Inequality holding for $\mathbb{P}_{\mathfrak{s}}$.

Note that for subadditive $\mathbb{B}$, the belief support probabilities still add for hypotheses of the same definiteness, and thus the support form corresponding to Tversky's and Koehler's theory apply to hypotheses that have the same v-value. This suggests the following three principles for subadditive $\mathbb{B}$:

1. The hypotheses naturally partition into families, with the hypotheses in each family having the same definiteness and hypotheses from different families having different definitenesses.

2. For each family, the support form of Tversky's and Koehler's theory holds for hypotheses from the family.

3. The support form of Rottenstreich's and Tversky's theory holds for hypotheses from different families.

7.6 Conclusions

The Choice Axioms yield a theory for a finitely additive conditional probability function (Theorem 7.4). Deleting the Axiom of Binary Symmetry from the Choice Axioms produces a generalization that yields a conditional belief function $\mathbb{B}$ that has the form,

$$\mathbb{B}(A|B) = \mathbb{P}(A|B)v(A)\,, \tag{7.18}$$

where $\mathbb{P}$ is a conditional probability function. $\mathbb{B}$ has limited applicability for modeling general conditional belief situations, because the definiteness of the focal event, $v(A)$ in Equation 7.18, does not depend on its conditioning event, B. However, there are situations in which Equation 7.18 does apply. One is Ellsberg Paradox as described in Subsection 7.4.2, and another is Support Theory as described in Section 7.5.

In support theory, each focal hypothesis A has a theoretical amount of unpackedness. The amount of unpackedness of A is a property of A and does not depend on the conditioning hypothesis B. In Section 7.5, $v(A)$ was interpreted as a multiplicative factor that accounts for the effects of unpackedness on the probability judgment.

In Chapter 10, Support Theory is formalized using an event space that is not a boolean algebra. Unpackedness and other non-Kolmogorov-like aspects of human probability judgments are given formulations in terms of this event space.

The next chapter presents a general way of describing non-boolean event spaces.

Chapter 8

Lattices

8.1 Introduction

One way of investigating generalizations of event spaces and the classical propositional calculus is through lattice theory. Lattices are formulated in terms of concepts corresponding to the sure event (or "true"), the null event (or "false" or "impossible"), ways of combining objects corresponding to union (or disjunction), intersection (or conjunction), and in cases of primary interest to this book, various concepts corresponding to complementation (or negation). Lattices are a much studied subject in mathematics, and various theorems about them are used in Chapter 9.

Boolean algebras of events—the event space of classical probability theory—is a special kind of lattice, called a *boolean lattice*. Example 8.1 below shows that the Classical propositional calculus can also be viewed as a boolean lattice. Section 8.3 presents a theorem due to M. H. Stone showing that boolean lattices are isomorphic to boolean algebras of events. When applied to the classical propositional calculus (considered as a boolean lattice), Stone's theorem shows that (classical) propositions are interpretable as events and visa versa. Thus Stone's theorem is important for establishing that the probability functions on propositions have the same theory as probability functions on events.

Chapters 9 and 10 employ a generalization of boolean algebras to model the logical structures generated by empirically determined propositions and used in making subjective probability judgments. The generalization is described in Section 8.4. It has the same kind of structure as the event space generated by a nonstandard logic known as "the intuitionistic propositional calculus."

8.2 Boolean Lattices

Definition 8.1 $\preceq$ is said to be *partial ordering on A* if and only if A is a nonempty set and the following three conditions hold for all a, b, and c in A:

(*i*) $a \preceq a$;

(*ii*) if $a \preceq b$ and $b \preceq a$, then $a = b$; and

(*iii*) if $a \preceq b$ and $b \preceq c$, then $a \preceq c$. □

Definition 8.2 $\langle A, \preceq, u, z\rangle$ is said to be a *lattice (with unit element u and zero element z)* if and only if

(*i*) $\preceq$ is a partial ordering on a;

(*ii*) for each a and b in A, there exists a unique c in A, called the *join* of a and b and denoted by $a \sqcup b$, such that $a \preceq a \sqcup b$, $b \preceq a \sqcup b$, and for all d in A,
$$\text{if } a \preceq d \text{ and } b \preceq d\,, \text{ then } a \sqcup b \preceq d\,;$$

(*iii*) for each a and b in A, there exists a unique element c in A, called the *meet* of a and b and denoted by $a \sqcap b$, such that $a \sqcap b \preceq a$, $a \sqcap b \preceq b$, and for all d in A,
$$\text{if } d \preceq a \text{ and } d \preceq b\,, \text{ then } d \preceq a \sqcap b\,;$$
and

(*iv*) for all a in A, $z \preceq a$ and $a \preceq u$. □

Note that the definition of "lattice" in Definition 8.2 assumes that the lattice has a maximal element u and a minimal element z. In the literature, "lattice" is often defined without the assumption of the existence of such elements.

Let $\langle A, \preceq, u, z\rangle$ be a lattice. Then it easily follows that $\sqcup$ and $\sqcap$ are commutative and associative operations on A.

Convention 8.1 A lattice $\langle A, \preceq, u, z\rangle$ is often written as
$$\langle A, \preceq, \sqcup, \sqcap, u, z\rangle\,,$$
and various concepts is often defined in terms of $\langle A, \preceq, \sqcup, \sqcap, u, z\rangle$ instead of $\langle A, \preceq, u, z\rangle$. □

Definition 8.3 $\mathfrak{B} = \langle B, \preceq', \sqcup', \sqcap', u', z'\rangle$ is said to be a *sublattice* of the lattice $\langle A, \preceq, \sqcup, \sqcap, u, z\rangle$ if and only if $\mathfrak{B}$ is a lattice and

$$B \subseteq A \quad \preceq' \subseteq \preceq, \ \sqcup' \subseteq \sqcup, \ \sqcap' \subseteq \sqcap, \ u' = u, \text{ and } z' = z. \quad \square$$

Definition 8.4 Let $\mathfrak{A} = \langle A, \preceq, \sqcup, \sqcap, u, z\rangle$ be a lattice. Then $\mathfrak{A}$ is said to be *distributive* if and only if for all a, b, and c in A,

$$a \sqcap (b \sqcup c) = (a \sqcap b) \sqcup (a \sqcap c). \quad \square$$

The following theorem is immediate.

Theorem 8.1 *Each sublattice of a distributive lattice is distributive.* $\square$

Distributive lattices play a central role in the book. The following two definitions characterize the kinds of lattices that arise from the event space of classical probability theory as well as from the propositional calculus of classical logic.

Definition 8.5 Let $\mathfrak{A} = \langle A, \preceq, \sqcup, \sqcap, u, z\rangle$ be a lattice and a be an element of A. Then an element b in A is said to be a *complement* of a if and only if

$$a \sqcap b = z \text{ and } a \sqcup b = u.$$

$\mathfrak{A}$ is said to be *complemented* if and only if a complement exists for each element in A. $\mathfrak{A}$ is said to be *uniquely complemented* if and only if each element in a has a unique complement. $\square$

Definition 8.6 A complemented distributive lattice is called a *boolean lattice.* $\square$

The notion of "complement" naturally generalizes into two weaker versions complement. One of these—$\sqcap$-complement (Definition 8.14 below)—provides for a generalization of boolean lattice that plays a major role in Chapters 9.

Theorem 8.2 *Each boolean lattice is uniquely complemented.*

Proof. Let $\mathfrak{A} = \langle A, \preceq, \sqcup, \sqcap, u, z\rangle$ be a boolean lattice and a be an arbitrary element of A. Suppose b and c are complements of a. Then

$$b = b \sqcap (a \sqcup c) = (b \sqcap a) \sqcup (b \sqcap c) = b \sqcap c,$$

showing $b \preceq c$. Similarly,

$$c = c \sqcap (a \sqcup b) = (c \sqcap a) \sqcup (c \sqcap b) = c \sqcap b,$$

showing $c \preceq b$. Thus, because $b \preceq c$ and $c \preceq b$, $b = c$. $\square$

Definition 8.7 Let $\mathfrak{A} = \langle A, \preceq, \sqcup, \sqcap, u, z\rangle$ be a boolean lattice and a be an arbitrary element of A. Then the unique complement of a described in Theorem 8.2 is denoted by $-a$. □

Convention 8.2 Usually boolean lattices $\mathfrak{A} = \langle A, \preceq, \sqcup, \sqcap, u, z\rangle$ will be described by the notation,

$$\mathfrak{A} = \langle A, \preceq, \sqcup, \sqcap, -, u, z\rangle,$$

explicitly listing the boolean complementation operation $-$. □

Example 8.1 (Boolean Lattice of Classical Propositions) The classical propositional calculus from logic is an important example of a boolean lattice. Let $\rightarrow$, $\leftrightarrow$, $\vee$, $\wedge$, and $\neg$ stand for, respectively, the logical connectives of implication ("if ... then"), logical equivalence, ("if and only if"), disjunction ("or"), conjunction, ("and"), and negation ("not"). Consider two propositions α and β to be equivalent if and only if $\alpha \leftrightarrow \beta$ is a tautology. Equivalent propositions partition the set of propositions into equivalence classes. Let A be the set of equivalence classes. The equivalence class containing a tautology is denoted by u, and the equivalence class containing a contradiction is denoted by z. By definition, equivalence class $a \preceq$ equivalence class b if and only if for some elements α in a and β in b, $\alpha \rightarrow \beta$ is a tautology. It is easy to check that $\preceq$ is a partial ordering on A.

Let a and b be arbitrary elements of A, α and β be respectively arbitrary elements of a and b, j be the equivalence class of $\alpha \vee \beta$, m be the equivalence class of $\alpha \wedge \beta$, and c be the equivalence class of $\neg\alpha$. Then it is easy to check that j is the $\preceq$-join of a and b, m is the $\preceq$-meet of a and b, c is the $\preceq$-complement of a, and

$$\mathfrak{A} = \langle A, \preceq, \sqcup, \sqcap, -, u, z\rangle,$$

is a boolean lattice. □

8.3 Representation Theorems of Stone

Definition 8.8 $\mathfrak{A} = \langle A, \preceq, \sqcup, \sqcap, u, z\rangle$ is said to be a *lattice of sets* if and only if $\mathfrak{A}$ is a lattice, u is a nonempty set, each element of A is a subset of u, $z = \varnothing$, $\preceq = \subseteq$, $\sqcup = \cup$, and $\sqcap = \cap$.

$\mathfrak{A}$ is said to be a *boolean lattice of sets* if and only if $\mathfrak{A}$ is a lattice of sets and $\mathfrak{A}$ is boolean. □

Note that it follows from properties of $\cap$ and $\cup$ that a lattice of sets is automatically distributive. Also note that for each boolean algebra of subsets $\mathcal{E}$ of X, that $\langle \mathcal{E}, \cup, \cap, -, X, \varnothing \rangle$ is a boolean lattice of sets.

M. H. Stone (1936, 1937) showed that every distributive lattice was isomorphic to a lattice of sets and that every boolean lattice was isomorphic to a boolean lattice of sets. The proofs given here of these theorems are based on Rasiowa and Sirkorski (1963) presentation of Stone (1937). In order to show Stone's theorems, some preliminary definitions and lemmas are needed.

Definition 8.9 Let $\mathfrak{A} = \langle A, \preceq, \sqcup, \sqcap, u, z \rangle$ be a lattice.

F is said to be a *filter (of $\mathfrak{A}$)* if and only if F is a nonempty subset of A and for all a and b in A,

(1) $z \notin F$.

(2) if $a \in F$ and $b \in F$, then $a \sqcap b \in F$, and

(3) if $a \in F$ and $a \preceq b$, then $b \in F$.

For each $a \neq z$ in A, $\{b \mid a \preceq b\}$ is called the *principal filter generated by a*. It easily follows that for each $a \neq z$ in A, the principal filter generated by a is a filter of $\mathfrak{A}$. □

Lemma 8.1 *Let $\mathfrak{A} = \langle A, \preceq, \sqcup, \sqcap, u, z \rangle$ be a lattice and $\mathcal{F}$ be a nonempty set of filters. Then $\bigcap \mathcal{F}$ is a filter.* □

Proof. It follows from the definition of "filter" that $u \in F$ for each $F \in \mathcal{F}$. Thus $\bigcap \mathcal{F} \neq \varnothing$. Then it immediate follows from the definition of "filter" that $\bigcap \mathcal{F}$ is a filter. □

Lemma 8.1 justifies the following definition:

Definition 8.10 Let $\mathfrak{A} = \langle A, \preceq, \sqcup, \sqcap, u, z \rangle$ be a lattice. For each nonempty subset B of A, $\bigcap \{F \mid F$ is a filter of $\mathfrak{A}$ and $B \subseteq F\}$ is said to be the *filter generated by B (of $\mathfrak{A}$)*. Using Lemma 8.1, it is easy to show that the filter generated by B is a filter of $\mathfrak{A}$. □

The following lemma is an immediate consequence of the above definitions:

Lemma 8.2 *Let $\mathfrak{A} = \langle A, \preceq, \sqcup, \sqcap, u, z \rangle$ be a lattice, $B \subseteq A$, and F be the filter generated by B. Then*

$$F = \{f \mid \textit{there exist } b_1, \ldots, b_n \textit{ in } B \textit{ and } b_1 \sqcap \cdots \sqcap b_n \preceq f\}.$$

Proof. Let G be an arbitrary filter such that $B \subseteq G$ and let

$$H = \{f \mid \text{there exist } b_1, \ldots, b_n \text{ in } B \text{ and } b_1 \sqcap \cdots \sqcap b_n \preceq f\}.$$

Then it follows from the definition of "filter" that $H \subseteq G$, and therefore by Definition 8.10 that $H \subseteq F$. It easily follows that H satisfies the definition of "filter" and $B \subseteq H$. Therefore $F \subseteq H$. Thus $F = H$. □

Definition 8.11 Let $\mathfrak{A} = \langle A, \preceq, \sqcup, \sqcap, u, z\rangle$ be a lattice. Then $\mathcal{F}$ is said to be a *chain of filters* of $\mathfrak{A}$ if and only if $\mathcal{F}$ is a nonempty set of filters such that for all F and G in $\mathcal{F}$, either $F \subseteq G$ or $G \subseteq F$. □

Lemma 8.3 *Let* $\mathfrak{A} = \langle A, \preceq, \sqcup, \sqcap, u, z\rangle$ *be a lattice and* $\mathcal{F}$ *be a chain of filters of* $\mathfrak{A}$. *Then* $\bigcup \mathcal{F}$ *is a filter of* $\mathfrak{A}$.

Proof. The proof is a straightforward verification and left to the reader. □

Definition 8.12 Let $\mathfrak{A} = \langle A, \preceq, \sqcup, \sqcap, u, z\rangle$ be a lattice. Then F is said to be an *ultrafilter* of $\mathfrak{A}$ if and only if F is a filter of $\mathfrak{A}$ and for each filter G of $\mathfrak{A}$, if $F \subseteq G$ then $F = G$. □

Lemma 8.4 *Let* $\mathfrak{A} = \langle A, \preceq, \sqcup, \sqcap, u, z\rangle$ *be a lattice. Then each filter of* $\mathfrak{A}$ *is contained in an ultrafilter of* $\mathfrak{A}$.

Proof. Let F be a filter on $\mathfrak{A}$ and

$$\mathcal{G} = \{G \mid G \text{ is a filter of } \mathfrak{A} \text{ and } F \subseteq G\}.$$

Then $\mathcal{G} \neq \varnothing$, because $F \in \mathcal{G}$, and if $\mathcal{H}$ is a chain of filters such that $\mathcal{H} \subseteq \mathcal{G}$, then $\bigcup \mathcal{H}$ is in $\mathcal{G}$. Thus by Zorn's Lemma (Definition 1.10), $\mathcal{G}$ has a $\subseteq$-maximal element, and this maximal element is an ultrafilter containing F. □

Definition 8.13 Let $\mathfrak{A} = \langle A, \preceq, \sqcup, \sqcap, u, z\rangle$ be a lattice. Then F is said to be a *prime* filter of $\mathfrak{A}$ if and only if F is a filter of $\mathfrak{A}$ and for each a and b in A,

$$\text{if } a \sqcup b \in F \text{ then } a \in F \text{ or } b \in F. \quad \square$$

Lemma 8.5 *Let* $\mathfrak{A} = \langle A, \preceq, \sqcup, \sqcap, u, z\rangle$ *be a distributive lattice. Then each ultrafilter of* $\mathfrak{A}$ *is prime.*

Proof. Suppose F is an ultrafilter of $\mathfrak{A}$ that is not prime. A contradiction will be shown. Let a and b be elements of A such that

$$a \sqcup b \in F,\ a \notin F, \text{ and } b \notin F.$$

Let

$$G = \{x \,|\, \text{there exists } c \in F \text{ such that } a \sqcap c \preceq x\}.$$

It easily follows that if d and e are in G, then $d \sqcap e$ is in G, and for all y in A, if $d \preceq y$, then y is in G. Thus to show that G is a filter, it needs only be shown that $z \notin G$. Suppose $z \in G$. A contradiction will be shown. By the definition of G, let c in F be such that $a \sqcap c = z$. Then, because $(a \sqcup b) \in F$ and $\mathfrak{A}$ is distributive, it follows that

$$c \sqcap (a \sqcup b) = (c \sqcap a) \sqcup (c \sqcap b) = z \sqcup (c \sqcap b) = c \sqcap b$$

is in F. Then, because $c \sqcap b \preceq b$, b is in F, contradicting $b \notin F$. Thus G is a filter. It is immediate that $F \subseteq G$. $a \in G$, because $u \in G$ and $a = a \sqcap u$. Thus, because by hypothesis $a \notin F$, $F \subset G$, contradicting that F is an ultrafilter. □

Lemma 8.6 *Let $\mathfrak{A} = \langle A, \preceq, \sqcup, \sqcap, u, z\rangle$ be a distributive lattice and a and b be elements of A such that not $b \preceq a$. Then there exists a prime filter F such that*

$$a \notin F \ \ \textit{and} \ \ b \in F.$$

Proof. Let $\mathcal{G}$ be the set of all filters G such that

$$a \notin G \ \text{ and } \ b \in G.$$

Then $\mathcal{G}$ is nonempty because it contains the principal filter generated by b. Using Lemma 8.3, it immediately follows that $\bigcup \mathcal{H}$ is in $\mathcal{G}$ for each chain of filters $\mathcal{H}$ of $\mathfrak{A}$ such that $\mathcal{H} \subseteq \mathcal{G}$. Thus by Zorn's Lemma, let H be a largest element of $\mathcal{G}$ in terms of the $\subseteq$-ordering. Then

$$a \notin H \ \text{ and } \ b \in H.$$

Thus to show the lemma, it needs only be shown that H is prime. This is done by contradiction. Suppose H is not prime. Let c and d be elements of A such that

$$c \sqcup d \in H,\ c \notin H, \ \text{ and } \ d \notin H.$$

Let

$$H_c = \{x \,|\, x \in A \text{ and there exists } h \text{ in } H \text{ such that } c \sqcap h \preceq x\}$$

and

$$H_d = \{x \,|\, x \in A \text{ and there exists } h \text{ in } H \text{ such that } d \sqcap h \preceq x\}.$$

We will show by contradiction that either $a \notin H_c$ or $a \notin H_d$:

Suppose $a \in H_c$ and $a \in H_d$. Then by the definitions of H_c and H_d, c_1 and d_1 in H can be found such that

$$c \sqcap c_1 \preceq a \text{ and } d \sqcap d_1 \preceq a\,.$$

Let $e = c_1 \sqcap d_1$. Then e is in H, and

$$c \sqcap e \preceq a \text{ and } d \sqcap e \preceq a\,.$$

Therefore,

$$a \succeq (c \sqcap e) \sqcup (d \sqcap e) = (c \sqcup d) \sqcap e \in H\,,$$

which implies $a \in H$, because $c \sqcup d$ and e are in H. This contradicts $a \notin H$.

Thus, without loss of generality suppose $a \notin H_c$. It is immediate from the definition of H_c that H_c satisfies all the conditions of a filter of $\mathfrak{A}$ except for, perhaps, the exclusion of the zero element of $\mathfrak{A}$, z, from being an element of H_c. $z \notin H_c$, because if it were, then a would be an element of H_c because $z \preceq a$, contradicting the above supposition that $a \notin H_c$. By the definition of H_c, $H \subseteq H_c$. It follows from the choice of H that $b \in H$, and thus $b \in H_c$. By the choice of c, $c \notin H$. However, because $c \in H_c$, it then follows that $H \subset H_c$. This together with the facts above that establish H_c is in $\mathcal{G}$ contradicts the choice of H as a $\subseteq$-largest element of $\mathcal{G}$. □

Theorem 8.3 (Representation Theorem for Distributive Lattices) *Let $\mathfrak{A} = \langle A, \preceq, \sqcup, \sqcap, u, z \rangle$ be a lattice. Then the following two statements are equivalent:*

1. *$\mathfrak{A}$ is distributive.*
2. *$\mathfrak{A}$ is isomorphic to a lattice of sets.*

Proof. Because each lattice of sets is distributive, it is immediate that Statement 2 implies Statement 1.

Suppose Statement 1. Let $\mathcal{P}$ be the set of prime filters of $\mathfrak{A}$. For each a in A, let

$$\varphi(a) = \{F \mid F \in \mathcal{P} \text{ and } a \in F\}\,,$$

and

$$\boldsymbol{\mathcal{P}} = \{\varphi(a) \mid a \in A\}.$$

It will be shown that φ is an isomorphism of $\langle A, \preceq, \sqcup, \sqcap, u, z \rangle$ onto $\langle \boldsymbol{\mathcal{P}}, \subseteq, \cup, \cap, \mathcal{P}, \varnothing \rangle$. Let a and b be arbitrary elements of A.

To show φ is one-to-one, suppose $a \neq b$. Then either not $a \preceq b$ or not $b \preceq a$. By Lemma 8.6, there is a prime filter F of $\mathfrak{A}$ such that F is in exactly one of the sets $\varphi(a)$, $\varphi(b)$. Therefore, $\varphi(a) \neq \varphi(b)$.

By the definition of $\mathcal{P}$, φ is onto $\mathcal{P}$.

Because the unit element u of $\mathfrak{A}$ is an element of each filter of $\mathfrak{A}$, it follows that $\varphi(u) = \mathcal{P}$, and because of the zero element z of $\mathfrak{A}$ is not an element of any filter of $\mathfrak{A}$, $\varphi(z) = \varnothing$.

Suppose F is an arbitrary element of $\varphi(a \sqcup b)$. Then F is a prime filter of $\mathfrak{A}$ and $(a \sqcup b) \in F$. Because F is prime, then either $a \in F$ or $b \in F$. Thus either $F \in \varphi(a)$ or $F \in \varphi(b)$, that is, $F \in (\varphi(a) \cup \varphi(b))$. This argument shows

$$\varphi(a \sqcup b) \subseteq \varphi(a) \cup \varphi(b)\,. \tag{8.1}$$

Suppose G is an arbitrary element of $\varphi(a) \cup \varphi(b)$. Then either $G \in \varphi(a)$ or $G \in \varphi(b)$, that is, either $a \in G$ or $b \in G$. Thus, because G is a filter, $(a \sqcup b) \in G$, and therefore, $G \in \varphi(a \sqcup b)$. This argument shows

$$\varphi(a) \cup \varphi(b) \subseteq \varphi(a \sqcup b)\,. \tag{8.2}$$

Thus, by Equations 8.1 and 8.2, $\varphi(a \sqcup b) = \varphi(a) \cup \varphi(b)$.

Suppose F is an arbitrary element of $\varphi(a \sqcap b)$. Then F is a prime filter of $\mathfrak{A}$ and $(a \sqcap b) \in F$. Then, because F is a filter of $\mathfrak{A}$ and $a \sqcap b \preceq a$ and $a \sqcap b \preceq b$, it follows that $a \in F$ and $b \in F$, that is, $F \in \varphi(a)$ and $F \in \varphi(b)$, and thus $F \in \varphi(a) \cap \varphi(b)$. This argument shows

$$\varphi(a \sqcap b) \subseteq \varphi(a) \cap \varphi(b)\,. \tag{8.3}$$

Suppose G is an arbitrary element of $\varphi(a) \cap \varphi(b)$. Then $G \in \varphi(a)$ and $G \in \varphi(b)$; that is, $a \in G$ and $b \in G$. Thus, because G is a filter, $a \sqcap b \in G$, and therefore $G \in \varphi(a \sqcap b)$. This argument shows

$$\varphi(a) \cap \varphi(b) \subseteq \varphi(a \sqcap b)\,. \tag{8.4}$$

Thus by Equations 8.3 and 8.4, $\varphi(a \sqcap b) = \varphi(a) \cap \varphi(b)$. □

Theorem 8.4 (Stone Representation Theorem) *Let* $\mathfrak{A} = \langle A, \preceq, \sqcup, \sqcap, u, z\rangle$ *be a lattice. Then the following two statements are equivalent:*

1. *$\mathfrak{A}$ is boolean.*

2. *$\mathfrak{A}$ is isomorphic to a boolean lattice of sets.*

Proof. It is immediate that Statement 2 implies Statement 1.

Suppose Statement 1. By Theorem 8.3, let $\mathfrak{X} = \langle \mathcal{E}, \subseteq, \cup, \cap, X, \varnothing\rangle$ be a lattice of sets and φ be a function such that φ is an isomorphism of $\mathfrak{A}$ onto $\mathfrak{X}$. Let x be an arbitrary element of X and $a = \varphi^{-1}(x)$. Then by isomorphism,

$$\varphi(a) \cup \varphi(\,-\,a) = \varphi(u) \;\text{ and }\; \varphi(a) \cap \varphi(\,-\,a) = \varphi(\varnothing)\,,$$

and thus

$$x \cup \varphi(-a) = X \text{ and } x \cap \varphi(-a) = \varnothing,$$

establishing that $\varphi(-a)$ is the complement of x. □

Probability functions often have as their domain a set of propositions (formulated in classical logic). Mathematicians and others often interpret such situations as probability functions on an event space, where the events form a boolean lattice of sets. For most probabilistic situations, this practice is justified by the Stone Representation Theorem.

8.4 $\sqcap$-Complemented Lattices

An axiomatic generalization of the Kolmogorov theory of probability can be achieved by generalizing the domains of probability to certain types of non-boolean lattices. The type of primary interest for the developments in the book is based on $\sqcap$-complementation.

Definition 8.14 Let $\mathfrak{A} = \langle A, \preceq, \sqcup, \sqcap, u, z\rangle$ be a lattice and a be an arbitrary element of A.

b in A is said to be the $\sqcap$-*complement of* a if and only if

(*i*) $b \sqcap a = z$, and

(*ii*) for all c in A, if $c \sqcap a = z$ then $c \preceq b$.

It easily follows that if the $\sqcap$-complement of a exists, then it is unique.

$\mathfrak{A}$ is said to be $\sqcap$-*complemented* if and only if each of its elements has a $\sqcap$-complement.

The $\sqcap$-complement of a, when it exists, is denoted by $\ominus A$. □

$\sqcap$-complementation is an important concept in lattice theory and logic, and it plays a central role in the book. The following related concept of "$\sqcup$-complement" plays a minor role in the book.

In the literature, a $\sqcap$-complement is generally called a *pseudo complement.*

Definition 8.15 The $\sqcup$-*complement,* c, of a is defined by

(i') $c \sqcup a = u$, and

(ii'') for all d in A, if $d \sqcup a = u$ then $d \succeq c$.

$\mathfrak{A}$ is said to be $\sqcup$-*complemented* if and only if each of its elements has a $\sqcup$-complement. It easily follows that the $\sqcup$-complement, when it exists, is unique. $\square$

Theorem 8.5 *Let $\mathfrak{A} = \langle A, \preceq, \sqcup, \sqcap, u, z\rangle$ be a distributive lattice, a an arbitrary element of A, b the $\sqcap$-complement of a, and c the $\sqcup$-complement of a. Then $b \preceq c$.*

Proof:

$$b = b \sqcap u = b \sqcap (a \sqcup c) = (b \sqcap a) \sqcup (b \sqcap c) = z \sqcup (b \sqcap c) = b \sqcap c \preceq c\,. \square$$

It is an immediate consequence of definitions of "complement," "$\sqcap$-complement," and "$\sqcup$-complement" that if the $\sqcap$-complement of a lattice $\mathfrak{A}$ = the $\sqcup$-complement of $\mathfrak{A}$, then $\mathfrak{A}$ is complemented.

8.4.1 Lattices of Open Sets

Because a $\sqcap$-complemented distributive lattice is distributive, it has a representation as a lattice $\mathfrak{X}$ of sets (Theorem 8.3). Theorem 8.8 below show that the sets in $\mathfrak{X}$ can be taken to be open sets in a topology with $\sqcap$-complementation being given a special topological interpretation. Topological representations are used to model scientific events in Chapter 9 and important cognitive processes involved in human probability judgments in Chapter 10.

Definition 8.16 A collection $\mathcal{U}$ is said to be a *topology with universe X* if and only if X is a nonempty set, $X \in \mathcal{U}$, $\varnothing \in \mathcal{U}$, for all A and B in $\mathcal{U}$, $A \cap B$ is in $\mathcal{U}$, and for all nonempty $\mathcal{Z}$ such that $\mathcal{Z} \subseteq \mathcal{U}$,

$$\bigcup \mathcal{Z} \text{ is in } \mathcal{U}\,. \tag{8.5}$$

Note that it is immediate from Equation 8.5 that for all A and B in $\mathcal{U}$, $A \cup B$ is in $\mathcal{U}$.

Let G be an arbitrary subset of X and $\mathcal{U}$ be a topology with universe X. Then the following two definitions hold:

- G is said to be *open (in the topology $\mathcal{U}$)* if and only if $G \in \mathcal{U}$.
- G is said to be *closed (in the topology $\mathcal{U}$)* if and only if $X - G$ is open.

It immediately follows that X and $\varnothing$ are closed as well as open. In some cases $\mathcal{U}$ may have X and $\varnothing$ as the only elements that are both open and closed, while in other cases $\mathcal{U}$ may have additional elements that are both open and closed, or even have all of its elements being both open and closed. The following three definitions hold for all $G \subseteq X$:

- The *closure* of G, $\mathsf{cl}(G)$, is, the smallest closed set C such that $G \subseteq C$; that is,
$$\mathsf{cl}(G) = \bigcap \{B|B \text{ is closed and } G \subseteq B\}.$$
- The *interior* of G, $\mathsf{int}(G)$, is the largest open set D such that $D \subseteq G$; that is,
$$\mathsf{int}(G) = \bigcup \{F|F \text{ is open and } F \subseteq G\}.$$
- The *boundary* of G, $\mathsf{bd}(G)$, is $\mathsf{cl}(G) - \mathsf{int}(G)$.

It easily follows that the definition of "topology" implies the existence of the closure, interior, and boundary of G for all $G \subseteq X$, and that for all closed F, $\mathsf{cl}(F) = F$. □

Theorem 8.6 *Suppose $\mathfrak{X} = \langle \mathcal{X}, \cup, \cap, -, X, \varnothing \rangle$ is a boolean lattice of sets. Let $\mathcal{U}$ = the set of all subsets of X. Then $\mathcal{U}$ is a topology in which each element of $\mathcal{U}$ is both open and closed. Thus in particular each element of $\mathcal{X}$ is both open and closed with respect to the topology $\mathcal{U}$.*

Proof. Let A be an arbitrary element of $\mathcal{U}$. Then $-A$ is closed. However, because $\mathsf{cl}(-A) = -A \in \mathcal{U}$, $-A$ is also open. □

Definition 8.17 $\mathfrak{X} = \langle \mathcal{X}, \subseteq, \cup, \cap, \ominus, X, \varnothing \rangle$ is said to be a *lattice of open sets of $\mathcal{U}$* if and only if $\langle \mathcal{X}, \subseteq, \cup, \cap, \ominus, X, \varnothing \rangle$ is a lattice of sets, $\mathcal{U}$ is a topology, $\mathcal{X} \subseteq \mathcal{U}$, and *with respect to $\mathcal{U}$*,
$$\ominus A = \mathsf{int}(\mathsf{cl}(-A)),$$
for all A in $\mathcal{X}$.

$\mathfrak{X} = \langle \mathcal{X}, \subseteq, \cup, \cap, \ominus, X, \varnothing \rangle$ is said to be a *lattice of open sets* if and only if for some $\mathcal{U}$, $\mathfrak{X}$ is a lattice of open sets of $\mathcal{U}$. □

Theorem 8.7 *Suppose $\mathfrak{X} = \langle \mathcal{X}, \subseteq, \cup, \cap, \ominus, X, \varnothing \rangle$ is a lattice of open sets. Then $\ominus$ is the $\cap$-complementation operation of $\mathfrak{X}$, and thus $\mathfrak{X}$ is a $\cap$-complemented set lattice (Definition 8.14).*

Proof. Immediate from Definitions 8.17 and 8.14. □

Example 8.2 (Topological Lattices of Open Sets) Let $\mathcal{U}$ be a topology with universe U. Then
$$\langle \mathcal{U}, \subseteq, \cup, \cap, \ominus, U, \varnothing \rangle$$
is a lattice of open sets. More generally,
$$\langle \mathcal{V}, \subseteq, \cup, \cap, \ominus, U, \varnothing \rangle$$
is a lattice of open sets, where $\mathcal{V}$ is a subset of $\mathcal{U}$ such that U and $\varnothing$ are in $\mathcal{V}$, and whenever E and F are in $\mathcal{U}$, $E \cup F$, $E \cap F$, and $\ominus E$ are in $\mathcal{V}$. □

Theorem 8.8 *Each $\sqcap$-complemented distributive lattice is isomorphic to a lattice of open sets (Definition 8.17).*

Proof. Let $\mathfrak{A} = \langle \mathcal{L}, \subseteq, \sqcup, \sqcap, \ominus_{\mathfrak{A}}, u, z\rangle$ be an arbitrary $\sqcap$-complemented distributive lattice. By Theorem 8.3, let φ be an isomorphism from $\mathfrak{L} = \langle \mathcal{L}, \subseteq, \sqcup, \sqcap, u, z\rangle$ onto a lattice of sets, $\langle \mathcal{X}, \subseteq, \cup, \cap, X, \varnothing\rangle$. Define the following operation $\ominus$ on $\mathcal{X}$ as follows: For all A in $\mathcal{X}$,

$$\ominus A = \varphi^{-1}(\ominus_{\mathfrak{A}} A)\,.$$

It then easily follows that

$$\mathfrak{X} = \langle \mathcal{X}, \subseteq, \cup, \cap, \ominus, X, \varnothing\rangle$$

is isomorphic to $\mathfrak{A}$ and is a $\mathfrak{X}$ is $\cap$-complemented lattice. Thus to show the theorem, we need to only find a topology $\mathcal{U}$ such that $\mathfrak{X}$ is a lattice of open sets of $\mathcal{U}$; that is, find a topology $\mathcal{U}$ such that $\ominus'$ defined on $\mathcal{X}$ by,

$$\text{for all } A \text{ in } \mathcal{X},\ \ominus' A = \mathsf{int}(\mathsf{cl}(-A)),$$

is the operation $\ominus$ of $\cap$-complementation for $\mathfrak{X}$.

Let

$$\mathcal{U} = \{\bigcup \mathcal{F} \,|\, \mathcal{F} \subseteq \mathcal{X}\}\,.$$

Then for each A in $\mathcal{X}$, $\bigcup\{A\} = A$ is in $\mathcal{U}$. Let $\varnothing \subseteq \mathcal{H} \subseteq \mathcal{U}$. Then for each H in $\mathcal{H}$, $H \in \mathcal{X}$, and thus by the definition of $\mathcal{U}$, $\bigcup \mathcal{H}$ is in $\mathcal{U}$. Because for subsets $\mathcal{F}$ and $\mathcal{G}$ of $\mathcal{X}$,

$$\bigcup \mathcal{F} \cap \bigcup \mathcal{G} = \bigcup\{F \cap G \,|\, F \in \mathcal{F} \text{ and } G \in \mathcal{G}\}\,,$$

it then easily follows from the above considerations that $\mathcal{U}$ is a topology and $\mathfrak{X}$ is an open set lattice of $\mathcal{U}$.

Let H be an arbitrary element of $\mathcal{X}$ and, with respect to $\mathcal{U}$, let

$$\ominus' H = \mathsf{int}(\mathsf{cl}(-H))\,.$$

Because H and $\ominus H$ are in $\mathcal{X}$, they are in $\mathcal{U}$, that is, H and $\ominus H$ are open. Therefore $-H$ is closed, and thus

$$\mathsf{cl}(-H) = -H\,.$$

Therefore, by the definition of $\ominus' H$, $H \cap \ominus' H = \varnothing$. Because $\ominus H \subseteq -H$ and $\ominus' H$ is the interior of $\mathsf{cl}(-H)$ and $\ominus H$ is open, it follows that $\ominus H \subseteq \ominus' H$. Thus to show the theorem, it needs to only be shown that

$\ominus' H \subseteq \ominus H$. This will be done by contradiction. Suppose x is in $\ominus' H$ but x is not in $\ominus H$. Because $\ominus' H$ is open, let $\mathcal{F} \subseteq \mathcal{X}$ be such that

$$\ominus' H = \bigcup \mathcal{F}.$$

Then because x is in $\ominus' H$, let F in $\mathcal{F}$ be such $x \in F$. Then $F \in \mathcal{X}$ and $F \subseteq \ominus' H$. Because $\ominus' H \subseteq -H$, $F \cap H = \varnothing$. Therefore, because $\ominus$ is the operation $\cap$-complementation for $\mathfrak{X}$, $F \subseteq \ominus H$, and therefore x is in $\ominus H$, a contradiction. $\square$

8.4.2 Properties of lattices of open sets

The following theorem shows properties involving the complementation operation $\ominus$ of a lattice of open sets that correspond to well-known properties of the complementation operation of a boolean lattice of sets.

Theorem 8.9 *Suppose $\mathfrak{X} = \langle \mathcal{X}, \subseteq, \cup, \cap, \ominus, X, \varnothing \rangle$ is a lattice of open sets. Then the following eight statements are true for all A and B in $\mathcal{X}$:*

1. $\ominus X = \varnothing$ *and* $\ominus \varnothing = X$.
2. *If* $A \cap B = \varnothing$, *then* $B \subseteq \ominus A$.
3. $A \cap \ominus A = \varnothing$.
4. *If* $B \subseteq A$, *then* $\ominus A \subseteq \ominus B$.
5. $A \subseteq \ominus \ominus A$.
6. $\ominus A = \ominus \ominus \ominus A$.
7. $\ominus (A \cup B) = \ominus A \cap \ominus B$.
8. $\ominus A \cup \ominus B \subseteq \ominus (A \cap B)$.

Proof. Statements 1 to 3 are immediate from Definition 8.14.

4. Suppose $B \subseteq A$. By Statement 3, $A \cap \ominus A = \varnothing$. Thus $B \cap \ominus A = \varnothing$. Therefore, by Statement 2, $\ominus A \subseteq \ominus B$.

5. By Statement 3, $\ominus A \cap A = \varnothing$. Thus by Statement 2, $A \subseteq \ominus \ominus A$, showing Statement 5.

6. By Statement 5, $A \subseteq \ominus \ominus A$. Thus by Statement 4,

$$\ominus \ominus \ominus A \subseteq \ominus A.$$

However, by Statement 5,

$$\ominus A \subseteq \ominus \ominus (\ominus A) = \ominus \ominus \ominus A.$$

Therefore, $\ominus A = \ominus \ominus \ominus A$.

7. By Statement 3, (i) $\ominus A \subseteq -A$ and (ii) $\ominus B \subseteq -B$. Thus

$$\ominus A \cap \ominus B \subseteq -A \cap -B = -(A \cup B).$$

Therefore $(A \cup B) \cap (\ominus A \cap \ominus B) = \varnothing$, and thus by Statement 2,

$$\ominus A \cap \ominus B \subseteq \ominus(A \cup B). \tag{8.6}$$

Because $A \subseteq A \cup B$ and $B \subseteq A \cup B$, it follows from Statement 4 that

$$\ominus(A \cup B) \subseteq \ominus A \quad \text{and} \quad \ominus(A \cup B) \subseteq \ominus B.$$

Therefore

$$\ominus(A \cup B) \subseteq \ominus A \cap \ominus B. \tag{8.7}$$

Equations 8.6 and 8.7 show that

$$\ominus(A \cup B) = \ominus A \cap \ominus B.$$

8. From

$$A \cap B \subseteq A \quad \text{and} \quad A \cap B \subseteq B,$$

it follows from Statement 4 that

$$\ominus A \subseteq \ominus(A \cap B) \quad \text{and} \quad \ominus B \subseteq \ominus(A \cap B),$$

and thus

$$\ominus A \cup \ominus B \subseteq \ominus(A \cap B). \quad \square$$

The following theorem gives some fundamental properties of $\ominus$ that differ from the corresponding properties of the complementation operation of boolean lattices of sets.

Theorem 8.10 *There exists a lattice of open sets $\mathfrak{X} = \langle \mathcal{X}, \subseteq, \cup, \cap, \ominus, X, \varnothing \rangle$ such that the following three statements are true about $\mathfrak{X}$.*

1. *For some A in $\mathcal{X}$, $A \cup \ominus A \neq X$.*
2. *For some A in $\mathcal{X}$, $\ominus \ominus A \neq A$.*
3. *For some A and B in $\mathcal{X}$, $\ominus(A \cap B) \neq \ominus A \cup \ominus B$.*

Proof. Let X be the set of real numbers, $\mathcal{X}$ be the usual topology on X determined by the usual ordering on X, C be the infinite open interval $(0, \infty)$, and D be the infinite open interval $(-\infty, 0)$. Then the reader can verify that Statement 1 follows by letting $A = C$, Statement 2 by letting $A = C \cup D$, and Statement 3 by letting $A = C$ and $B = D$. $\square$

8.5 Intuitionistic Propositional Logic

The Dutch mathematician L. L. J. Brouwer introduced intuitionism as an alternative form of mathematics. In intuitionism, mathematics is viewed as a form of mental activity in which the mathematician constructs mathematical objects through pure intuition about time, and in the process experiences the correctness of the construction and the truth of the proposition associated with it. Brouwer concluded that such intuitionistic methods of derivation were in principle non-axiomatizable. However, the published forms of the theorems he derived exhibited in their proofs enough regularity that an axiomatic approach to the proofs seemed feasible, and Heyting (1930) produced an axiomatic system for accomplishing this. Today logics that satisfy equivalents of Heyting's axiomatization are generically called *intuitionistic logic.* The formal structure of Heyting's formulation applies to other situations than a logic for Brouwer's intuitionism. For example, Kolmogorov (1932) showed that an equivalent of the intuitionistic propositional calculus had the correct formal properties for a theory for mathematical constructions, and Gödel (1933) showed that a different equivalent provided a logical foundation for proof theory of mathematical logic. In obtaining these results, Kolmogorov and Gödel provided interpretations to the logical primitives that were different from Heyting's. In Chapter 9, the intuitionistic versions of conjunction, disjunction, and negation are provided with interpretations that yield event spaces appropriate for the modeling of probabilistic empirical events.

In a manner similar to Example 8.1 (distributive lattice of propositions), the intuitionistic version of "if and only if" produces an equivalence relation on intuitionistic propositions, and the induced algebra on its equivalence classes yields a lattice, called a *pseudo boolean algebra.* Formally, it is defined as follows.

Definition 8.18 $\mathfrak{X} = \langle \mathcal{X}, \subseteq, \cup, \cap, \Rightarrow, X, \varnothing \rangle$, where $\Rightarrow$ is a binary operation on $\mathcal{X}$, is said to be a *pseudo boolean algebra of subsets* if and only if the following three conditions hold for all A and B in $\mathcal{X}$:

(*i*) $\mathfrak{X} = \langle \mathcal{X}, \subseteq, \cup, \cap, X, \varnothing \rangle$ is a lattice of sets.

(*ii*) $A \cap (A \Rightarrow B) \subseteq B$.

(*iii*) For all C in $\mathcal{X}$, if $A \cap C \subseteq B$ then $C \subseteq (A \Rightarrow B)$.

$\Rightarrow$ is called the operation of *relative pseudo complementation.* □

Let $\mathfrak{B} = \langle \mathcal{B}, \subseteq, \cup, \cap, -, X, \varnothing \rangle$ be a boolean algebra of sets. Then it is easily follows that

$$\langle \mathcal{B}, \subseteq, \cup, \cap, \Rightarrow, X, \varnothing \rangle$$

is pseudo boolean algebra of subsets, where for all A and B in $\mathcal{B}$,

$$A \Rightarrow B = (-A) \cup B .$$

Thus each boolean algebra of sets is a pseudo boolean algebra of subsets.

The implication operation $\rightarrow$ of classical logic corresponds to the expression $(-A) \cup B$ in the boolean algebra of sets. Similarly, the implication operation of intuitionistic logic corresponds to $\Rightarrow$ in the pseudo boolean algebra of subsets. In a pseudo boolean algebra of sets, the negation operator $\neg$ has the following definition:

$$\neg A = A \Rightarrow \varnothing .$$

Let $\langle \mathcal{X}, \subseteq, \cup, \cap, \neg, \Rightarrow, \varnothing \rangle$ be a pseudo boolean algebra of subsets and $\mathfrak{X} = \langle \mathcal{X}, \subseteq, \cup, \cap, \neg, X, \varnothing \rangle$. It is not difficult to verify that $\mathfrak{X}$ is a lattice of open sets with $\neg = \ominus$. In terms of the corresponding intuitionistic logic, $\neg$ $(= \ominus)$ corresponds to the negation operator of the logic.

Lattices of open sets are lattices of sets with a weakened form of complementation that has the fundamental properties of intuitionistic negation operation on propositions. Lattices of open sets and pseudo boolean algebras differ in that lattices of open sets assume the existence of a $\cap$-complement for each element of the domain X, whereas pseudo boolean algebras assume for all elements A and B in the lattice's domain, the existence of the pseudo complement A relative to B, $A \Rightarrow B$. However, besides the difference in primitives, $\ominus$ versus $\Rightarrow$, there is another difference that appears when axiomatizing the corresponding general lattice versions—as oppose to the lattice of sets versions—of the two kinds lattices. This is because the lattice of sets versions are automatically distributive. In the general lattice $\sqcap$-complementation version, it needs to be assumed that the lattice is distributive and that the $\sqcap$-complement exists for each element of the domain. In the general lattice version based on relative pseudo complementation, only the existence of relative pseudo complements for all pairs of elements of the domain needs to be assumed: the lattice being distributive then follows as a theorem.

8.6 Non-Boolean Event Spaces

Many results in lattice theory show that unique complementation with a little other structure often implies distributivity. Saliĭ (1988), who presents many theorems characterizing boolean lattices in terms of unique complementation, comments the following about this:

> The results of the preceding section cast great doubt on the existence of nondistributive uniquely complemented lattices. At the end of the 1930's the doubt (with a much smaller set of corroborating facts) gained widespread conviction. Thus the appearance in 1945 of [Dilworth's Theorem] was completely unexpected. *(p. 51)*

Dilworth's Theorem is formulated as follows:

Theorem 8.11 (Dilworth's Theorem) *Any lattice can be embedded in some uniquely complemented lattice.*

Proof. See Saliĭ (1988) p. 51. □

Dilworth's proof and subsequent proofs of his theorem did not produce an explicit example of a non-distributive uniquely complemented lattice. Saliĭ (1988) comments (his italics),

> *We know that nondistributive uniquely complemented lattices exist, but at present we do not have a single explicit example.* Such lattices have not yet been encountered in mathematical practice.

Thus, *as a practical matter*, the following working hypothesis is reasonable for the modeling of empirical phenomena: *All uniquely complemented lattices are boolean.* Therefore, in looking for alternative event spaces to boolean lattices, we can narrow our considerations to the following:

(1) Non-complemented lattices.

(2) Complemented but not uniquely complemented lattices.

(3) Event spaces that are not lattices.

Strategy (1) was pursued in this chapter and is used in Chapter 9 to model a probability theory for verifiable or refutable scientific events. (2) is employed in quantum mechanics and its mathematical generalizations—topics not covered in this book. And (3) is pursued in Chapter 10 to model human judgments of probability.

Chapter 9

Belief for Scientific Events and Refutations

9.1 Introduction

This chapter follows on Narens (2005) development of a probability calculus for empirical science. It axiomatizes scientific events as the elements of a $\cap$-complemented lattice of sets. This lattice is a part of a larger boolean lattice that in general has nonscientific elements—that is, has events that are neither verifiable or refutable. A probability function $\mathbb{P}$ is assumed to measure uncertainty on the larger boolean lattice. The restriction of $\mathbb{P}$ to the sublattice of scientific events then produces a probability calculus for scientific events with its own non-boolean "logic" of a $\cap$-complemented event space. Narens (2005) likens this approach to scientific probability event spaces to a commonly used approach in differential geometry:

> An analogous situation occurred in mathematics. Once, Euclidean geometry was completely dominate. Manifolds were described and worked out in a Euclidean framework. Later, it was found to be far more insightful and productive to work completely "inside" the manifold, exploiting its intrinsic characteristics. Thus, for example, it became more productive for mathematicians to describe the geometry of a torus (the surface of a doughnut shaped object) intrinsically in terms of its 2-dimensional geometrical properties than extrinsically as a 2-dimensional surface in a 3-dimensional Euclidean space. In the present case, the Kolmogorov theory is dominate. The scientific events, i.e., events of the form, "A occurs and is (scientifically) verifiable," or of the form, "A occurs and is (scientifically)

> refutable," of a targeted fragment of science can be viewed as a portion of a larger set of platonic events. They only form a portion, because the complement (or negation) of a scientific event need not be a scientific event; that is, the complement of a verifiable event need be neither verifiable nor refutable, and similarly, the complement of a disjunction of refutable events need be neither verifiable nor refutable. In short, the complementation operator of a boolean algebra of events does not in general preserve scientific events. In this sense, the Kolmogorov theory adds extrinsic structure to the scientific events through its use of (boolean) complementation.

9.2 Lattices of Scientific Events

Let $\mathfrak{S} = \langle \mathcal{S}, \subseteq, \cup, \cap, U, \varnothing \rangle$ be a lattice of sets. In the intended interpretation, $\mathcal{S}$ is a set of events about a phenomenon under investigation. It is the basis for the generation of additional events. A new operation $\ominus$ is introduced for showing that an event has been refuted. The refutation of an event Q in $\mathcal{S}$, $\ominus Q$, is an event that may be outside of $\mathcal{S}$, and $\ominus$ may be applied to $\ominus Q$ to produce the event that is the refutation of the refutation of Q, $\ominus(\ominus Q)$. Subsets E of U are considered to be events about the *phenomenon* under consideration. Some refutations may not be events about the phenomenon under consideration. In such situations, the refutations are considered to be about the *science of the phenomenon* under consideration. The new operation $\ominus$ is called *refutation.*

Convention 9.1 As usual, $\ominus\ominus A$ stands for $\ominus(\ominus A)$, $\ominus A \cup B$ for $(\ominus A) \cup B$, etc. □

Definition 9.1 Let Q be an event. Then, by definition, $\ominus Q$ is the event that the assumption of the occurrence of the event Q leads to a contradiction. "$\ominus Q$" is read as, "The event that the event Q has been refuted." □

Obviously Definition 9.1 is incomplete, because "leads to a contradiction" has not been specified. Fortunately, for the purposes of this chapter only a few basic properties of $\ominus$ are needed. Narens (2005) comments,

> Methods that produce scientific contradictions are many and varied. For example, an event Q may be refuted through verification of an event T such that $Q \cap T = \varnothing$. Or the assumption of the occurrence of Q may contradict a fundamental principle of the portion of science under consideration, e.g., in a portion

> of classical physics where the assumption of the occurrence of Q implies the existence of perpetual motion. For most rich fragments of science, we do not have complete descriptions of the methods of scientific inference. I consider it likely that such complete descriptions are incapable of formal description.

Definition 9.2 Let $\mathfrak{S} = \langle \mathcal{S}, \subseteq, \cup, \cap, U, \varnothing \rangle$ be a lattice of sets. Then

$$\mathfrak{E} = \langle \mathcal{E}, \subseteq, \cup, \cap, \ominus, X, \varnothing \rangle$$

is said to be the *lattice of scientific events (generated by $\mathcal{S}$)* if and only if $\mathcal{E}$ is the smallest set of events such that

(*i*) $\mathcal{S} \subseteq \mathcal{E}$;

(*ii*) $\varnothing$ and X are in $\mathcal{E}$;

(*iii*) if A is in $\mathcal{E}$, then $A \subseteq X$; and

(*iv*) if A and B are in $\mathcal{E}$, then $A \cup B$, $A \cap B$, and $\ominus A$ are in $\mathcal{E}$;

and the following three axioms about $\ominus$ hold:

Axiom 9.1 $\ominus X = \varnothing$ *and* $\ominus \varnothing = X$. □

Axiom 9.2 *For all A and B in $\mathcal{E}$, if $A \cap B = \varnothing$, then $B \subseteq \ominus A$.* □

Axiom 9.3 *For all A in $\mathcal{E}$, $A \cap \ominus A = \varnothing$.* □

Note that X is the sure event of $\mathcal{E}$, that is, X is the largest event in $\mathcal{E}$. In general, X is different from U, the largest event in $\mathcal{S}$.

Theorem 9.1 *Let $\mathfrak{E} = \langle \mathcal{E}, \subseteq, \cup, \cap, \ominus, X, \varnothing \rangle$ be a lattice of scientific events generated by $\mathcal{S}$. Then $\mathfrak{E}$ is a $\cap$-complemented lattice of sets (Definition 8.14).*

Proof. Let A be an arbitrary element of $\mathcal{E}$. It immediately follows from Axioms 9.2 and 9.3 that $\mathfrak{E}$ is a $\cap$-complemented lattice of sets. □

Definition 9.3 Let $\mathfrak{E} = \langle \mathcal{E}, \subseteq, \cup, \cap, \ominus, X, \varnothing \rangle$ be a lattice of scientific events generated by $\mathcal{S}$. Then $\mathcal{S}$ is called the set of *initial scientific events (of $\mathfrak{E}$)* and $\mathcal{E}$ is called the set of *scientific events (of $\mathfrak{E}$)*. Events of $\mathcal{E}$ of the form $\ominus A$ are called *refutations.*

$\mathfrak{F}$ is said to be a *lattice of scientific events* if and only if for some $\mathcal{T}$, $\mathfrak{F}$ is the lattice of scientific events generated by $\mathcal{T}$. □

The formal theory of scientific events presented here has applications outside of science, which is only natural because it is based on very general properties of contradictions contained in Axioms 9.1 to 9.3. By Theorem 9.1 and the discussion in Chapter 8 concerning intuitionistic logic, it therefore also applies to situations involving applications of intuitionistic logic. One example of this is the logic inherent in proof theory.

Example 9.1 (Proof Theory) Let Γ be formal theory of arithmetic used by Gödel (1931) in his famous incompleteness theorem, and let $\Gamma \vdash \alpha$ stand for the first-order sentence α is a formal consequence of Γ. Let $\equiv$ be the equivalence relation defined on the set of first-order arithmetic sentences, $\mathbf{F}$, such that for all ϕ and ψ in $\mathbf{F}$, $\phi \equiv \psi$ if and only if ($\Gamma \vdash \phi$ if and only if $\Gamma \vdash \psi$). Let $\mathcal{S}$ be the set of $\equiv$-equivalence classes. Define $\preceq$, $\vee$, $\wedge$, u, and z as follows: For all A and B in $\mathcal{S}$,

(*i*) $A \preceq B$ if and only if for some α in A and β in B, if $\Gamma \vdash \alpha$, then $\Gamma \vdash \beta$,

(*ii*) $A \vee B$ if and only if for some α in A and β in B, $\Gamma \vdash \alpha$ or $\Gamma \vdash \beta$,

(*iii*) $A \wedge B$ if and only if for some α in A and β in B, $\Gamma \vdash \alpha$ and $\Gamma \vdash \beta$,

(*iv*) u is the set of α in $\mathbf{F}$ such that $\Gamma \vdash \alpha$.

(*v*) z is the set of α in $\mathbf{F}$ such that $\Gamma \vdash \neg\, \alpha$, where $\neg$ is the negation operator of the language of Γ.

Then $\langle \mathcal{S}, \preceq, \vee, \wedge, u, z\rangle$ is a distributive lattice. For A in $\mathcal{S}$, let $\ominus A$ stand for, "For some α in A, $\Gamma \vdash \alpha$ leads to a contradiction." Let γ be the critical sentence ("I am not a theorem") that Gödel used to establish his incompleteness theorem, and let G be the element of $\mathcal{S}$ to which γ belongs. Gödel showed that the assumption of $\Gamma \vdash \gamma$ contradicted a fundamental principle of the metatheory of arithmetic, ω-consistency[1], which in the current setup yields $\ominus G$. $\square$

Note the following three analogies between refutation in science and refutation in this mathematical example:

1. Verifiability is analogous to provability from Γ.
2. Refuting A by verifying a B such that $A \cap B = \varnothing$ is analogous to assuming $\Gamma \cup \{\delta\}$ and showing $\Gamma \cup \{\delta\} \vdash \kappa$, where κ is a contradiction of $\mathbf{F}$, to conclude $\ominus D$, where D is the element of $\mathcal{S}$ to which δ belongs.

[1] *ω-consistency* is said to hold if and only if there is no formula $\phi(x)$ in the formal language describing arithmetic such that $\Gamma \vdash \phi(\mathbf{n})$ for each positive n and $\Gamma \vdash \exists x(\neg\, \phi(x))$. Gödel showed $\Gamma \vdash \gamma(\mathbf{n})$ for each positive integer n and $\Gamma \vdash \exists x(\neg\, \gamma(x))$.

3. Assuming the occurrence of A to contradict a fundamental principle of the portion of science under consideration is analogous to assuming $\Gamma \vdash \theta$ and showing that a fundamental principle of the metatheory of arithmetic is contradicted.

9.3 Boolean Lattice of Refutations

Definition 9.4 Let $\mathfrak{E} = \langle \mathcal{E}, \subseteq, \cup, \cap, \ominus, X, \varnothing \rangle$ be a lattice of scientific events. By definition, let

$$\mathcal{R} = \{\ominus B | B \in \mathcal{E}\} = \{A | A \text{ is a refutation in } \mathcal{E}\}.$$

By definition, for each A and B in $\mathcal{R}$, let

$$A \Cup B = \ominus \ominus (A \cup B).$$

Then $\mathfrak{R}$ is said to be the *boolean lattice of refutations (of $\mathfrak{E}$)* if and only if

$$\mathfrak{R} = \langle \mathcal{R}, \subseteq, \Cup, \cap, \ominus, X, \varnothing \rangle. \quad \square$$

Let $\mathfrak{E}$ be a lattice of scientific events and $\mathfrak{R}$ be the boolean lattice of refutations in $\mathfrak{E}$. The following lemmas and theorem establish facts about $\mathfrak{R}$, including that $\mathfrak{R}$ is a boolean lattice.

Hypotheses for Lemmas 9.1 to 9.6: Let $\mathfrak{E} = \langle \mathcal{E}, \subseteq, \cup, \cap, \ominus, X, \varnothing \rangle$ be a lattice of scientific events and $\mathfrak{R} = \langle \mathcal{R}, \subseteq, \Cup, \cap, \ominus, X, \varnothing \rangle$ be the boolean lattice of refutations of $\mathfrak{E}$. Let *A, B, and C be arbitrary elements of $\mathcal{R}$, and A' and B' be elements of $\mathcal{E}$ such that $A = \ominus A'$ and $B = \ominus B'$.*

Lemma 9.1 $A = \ominus \ominus A$.

Proof. Because $A = \ominus A'$, it follows from Statement 6 of Theorem 8.9 that

$$\ominus \ominus A = \ominus \ominus \ominus A' = \ominus A' = A. \quad \square$$

Lemma 9.2 *$A \cap B$ is in $\mathfrak{R}$.*

Proof. By Statement 7 of Theorem 8.9,

$$A \cap B = \ominus A' \cap \ominus B' = \ominus (A' \cup B'). \quad \square$$

Lemma 9.3 *If $B \subseteq A$, then $\ominus \ominus B \subseteq A$.*

Proof. By Lemma 9.1, $B = \ominus \ominus B$. $\square$

Lemma 9.4 $\mathfrak{R} = \langle \mathcal{R}, \subseteq, X, \varnothing \rangle$ *is a lattice with* $A \cap B$ *as the meet of* A *and* B, *and* $A \Cup B$ *as the join of* A *and* B.

Proof. Because $X = \neg \varnothing$ and $\varnothing = \neg X$, X and $\varnothing$ are in $\mathcal{R}$. Because X and $\varnothing$ are, respectively, the $\subseteq$-maximal and $\subseteq$-minimal elements of $\mathcal{E}$, they are, respectively, the $\subseteq$-maximal and $\subseteq$-minimal elements of $\mathcal{E}$.

By Lemma 9.2, $A \cap B$ is in $\mathcal{R}$. Because $A \cap B$ is the meet in $\mathfrak{E}$ and $\mathcal{R} \subseteq \mathcal{E}$, it follows that $A \cap B$ is the meet in $\mathfrak{R}$.

Let D be an arbitrary element in $\mathcal{R}$ such that

$$A \subseteq D \text{ and } B \subseteq D.$$

(Such a D exists, because the above expression holds for $D = X$.) To show that $A \Cup B$ is the join of A and B in $\mathfrak{R}$, it needs only to be shown that $A \subseteq A \Cup B \subseteq D$ and $B \subseteq A \Cup B \subseteq D$. By the choice of D, $A \cup B \subseteq D$. By two applications of Statement 4 of Theorem 8.9,

$$\ominus \ominus (A \cup B) \subseteq \ominus \ominus D.$$

By Statement 5 of Theorem 8.9, $A \cup B \subseteq \ominus \ominus (A \cup B)$, and by Lemma 9.1, $\ominus \ominus D = D$. Thus,

$$A \cup B \subseteq \ominus \ominus (A \cup B) = A \Cup B \subseteq D.$$

Therefore,

$$A \subseteq A \Cup B \subseteq D \text{ and } B \subseteq A \Cup B \subseteq D.$$

Thus $A \Cup B$ is the join of A and B in $\mathfrak{R}$. □

Lemma 9.5 $(A \cap B) \Cup (A \cap C) = A \cap (B \Cup C)$.

Proof.

$$\begin{aligned}
(A \cap B) \Cup (A \cap C) &= \ominus \ominus [(A \cap B) \cup (A \cap C)] \\
&= \ominus \ominus [A \cap (B \cup C)] \\
&\supseteq \ominus [\ominus A \cup \ominus (B \cup C)] && \text{(Statement 8 of Theorem 8.9)} \\
&= [\ominus \ominus A \cap \ominus \ominus (B \cup C)] && \text{(Statement 7 of Theorem 8.9)} \\
&= [\ominus \ominus A \cap (B \Cup C)] \\
&= A \cap (B \Cup C)] \quad \text{(Lemma 9.1)},
\end{aligned}$$

that is,

$$(A \cap B) \Cup (A \cap C) \supseteq A \cap (B \Cup C) \,. \tag{9.1}$$

Because,

$$\begin{aligned}(A \cap B) \cup (A \cap C) &= A \cap (B \cup C) \\ &\subseteq A \cap \ominus\ominus(B \cup C) \quad \text{(Statement 5 of Theorem 8.9)} \\ &= A \cap (B \Cup C),\end{aligned}$$

it follows that

$$(A \cap B) \cup (A \cap C) \subseteq A \cap (B \Cup C) \,.$$

Then

$$A \cap B \subseteq A \cap (B \Cup C) \ \text{ and } \ A \cap C \subseteq A \cap (B \Cup C) \,.$$

Thus, because by Lemma 9.4 $(A \cap B) \Cup (A \cap C)$ is the $\subseteq$-smallest element in $\mathcal{R}$ such that

$$A \cap B \subseteq (A \cap B) \Cup (A \cap C) \ \text{ and } \ A \cap C \subseteq (A \cap B) \Cup (A \cap C) \,,$$

it then follows that

$$(A \cap B) \Cup (A \cap C) \subseteq A \cap (B \Cup C). \tag{9.2}$$

Equations 9.1 and 9.2 show

$$(A \cap B) \Cup (A \cap C) = A \cap (B \Cup C) \,. \quad \square$$

Lemma 9.6 $A \cup \ominus A = X$.

Proof.

$$\begin{aligned}A \cup \ominus A &= \ominus\ominus(A \cup \ominus A) \quad \text{(Lemma 9.1)} \\ &= \ominus(\ominus A \cap \ominus\ominus A) \quad \text{(Statement 7 of Theorem 8.9)} \\ &= \ominus(\ominus A \cap A) \quad \text{(Lemma 9.1)} \\ &= \ominus\varnothing \quad \text{(Axiom 9.3)} \\ &= X. \quad \text{(Axiom 9.1)} \quad \square\end{aligned}$$

Theorem 9.2 $\mathfrak{R} = \langle \mathcal{R}, \subseteq, \Cup, \cap, \ominus, X, \varnothing \rangle$ *is a boolean lattice.*

Proof. $\subseteq$ partially orders $\mathcal{R}$, and X and $\varnothing$ are respectively the maximal and minimal elements of $\mathcal{R}$ with respect to $\subseteq$. By Lemma 9.4, $\cap$ and $\Cup$ are respectively the meet and join operations of $\mathfrak{R}$. By Lemma 9.5, $\cap$ distributes over $\Cup$, and by Lemma 9.6 and Axiom 9.3, $\ominus$ is the complement operator on $\mathcal{R}$. Thus $\mathfrak{R}$ is a boolean lattice. $\square$

In this section, scientific events will be viewed as events as part of boolean lattice of events. Because some scientific events do not have boolean complements that are scientific, it follows that some events in the boolean lattice are non-scientific. There are a number of ways of viewing such events. In this section they are viewed as events describing reality and are referred to "platonic events." Scientific events are considered here to be platonic events, for example, the refutation of an event Q, $\ominus Q$, is the platonic event of the refutation of Q occurring.

Definition 9.5 Let $\mathfrak{E} = \langle \mathcal{E}, \subseteq, \cup, \cap, \ominus, X, \varnothing \rangle$ be a lattice of scientific events. Let $\mathcal{B}$ be the smallest set such that

(*i*) $\mathcal{E} \subseteq \mathcal{B}$, and

(*ii*) for all A and B in $\mathcal{E}$, $A \cup B$, $A \cap B$, and $X - A$ are in $\mathcal{B}$.

For each C in $\mathcal{B}$, let $-C$ stand for $X - C$. Then

$$\mathfrak{B} = \langle \mathcal{B}, \subseteq, \cup, \cap, -, X, \varnothing \rangle$$

is called the *outer boolean lattice generated by* $\mathfrak{E}$. □

It is immediate that the outer boolean lattice generated by $\mathfrak{E}$ is a boolean lattice.

9.4 Probability Theory for Scientific Events

Many researchers believe that all rational approaches to probability theory will end up satisfying the Kolmogorov axioms. In this section, a boolean lattice of sets and a non-additive belief function on it are constructed, and it is argued that the non-additive belief function is a rational assignment degrees of belief to uncertainty *if the Kolmogorov theory is applicable to the kind of uncertainty generated by the outer boolean lattice of a lattice of scientific events.* It is my understanding of the literature that those who propose that rationality demands that belief uncertainty be measured in a way that satisfies the Kolmogorov axioms would agree, at least in theory, that a subjective probability function exists on the outer boolean algebra.

Because Kolmogorov defined probability in terms of the sure event, the empty event, set-theoretic union, and set-theoretic intersection, but not in terms of set-theoretic complementation, his definition applies to general lattices:

Definition 9.6 Let $\mathfrak{A} = \langle A, \preceq, \sqcup, \sqcap, u, z \rangle$ be a lattice. A *probability function* $\mathbb{P}$ on $\mathfrak{A}$ has the formal properties of a finitely additive measure, that is,

(*i*) $\mathbb{P}(u) = 1$ and $\mathbb{P}(z) = 0$;

(*ii*) for all a in A, $0 \leq \mathbb{P}(a) \leq 1$; and

(*iii*) for all a and b in A, if $a \sqcap b = z$, then $\mathbb{P}(a \sqcup b) = \mathbb{P}(a) + \mathbb{P}(b)$. □

Suppose $\mathfrak{E}$ is a lattice of scientific events and $\mathfrak{B}$ is the outer boolean lattice generated by $\mathfrak{E}$ and suppose rationality demands that uncertainty in $\mathfrak{B}$ be measured by a probability function $\mathbb{P}$. Then the restriction of $\mathbb{P}$ to the domain of $\mathfrak{E}$, $\mathbb{P}_{\mathcal{E}}$, is a probability function on the lattice $\mathfrak{E}$ (Definition 9.6).

Let $\mathfrak{R} = \langle \mathcal{R}, \subseteq, \Cup, \cap, \ominus, X, \varnothing \rangle$ be the boolean lattice of refutations of $\mathfrak{E}$ (Definition 9.4). Let $\mathbb{B}$ be the restriction of $\mathbb{P}_{\mathcal{E}}$ to $\mathcal{R}$. Then $\mathbb{B}$ may be *superadditive;* that is, for all A and B in $\mathcal{R}$ such that $A \cap B = \varnothing$,

$$\mathbb{B}(A) + \mathbb{B}(B) \leq \mathbb{B}(A \Cup B),$$

and there may be elements C and D in $\mathcal{R}$ such that $C \cap D = \varnothing$ and

$$\mathbb{B}(C) + \mathbb{B}(D) < \mathbb{B}(C \Cup D).$$

The superadditivity of $\mathbb{B}$ results from the fact that $\mathcal{R}$ may have elements A and B such that $A \cap B = \varnothing$ and $A \cup B$ is not a refutation, and therefore,

$$A \Cup B = \ominus\ominus (A \cup B) \supset A \cup B.$$

Let $\mathfrak{L} = \langle A, \leq, \sqcup, \sqcap, u, z \rangle$ be a lattice and $\preceq^*$ be a binary relation on A. Interpret "$a \preceq^* b$" as "the degree of belief for a is less than or equal to the degree of belief for b." For boolean lattices, the following condition has been considered as necessary for rationality:

Definition 9.7 Let $\mathfrak{L} = \langle A, \leq, \sqcup, \sqcap, u, z \rangle$ and $\preceq^*$ be as just above. Then $\preceq^*$ is said to be $\sqcup$-*monotonic* if and only if $\preceq^*$ is a total ordering on A and for all a, b, and c in A, if $a \sqcap c = z$ and $b \sqcap c = z$, then

$$a \prec^* b \text{ iff } a \sqcup c \prec^* b \sqcup c. \quad \square$$

Suppose $\mathfrak{L}$ above is boolean. In the behavioral sciences, "rationality" is formulated in terms of an individual's behavior. The reasons an individual may give for his or her behavior are generally ignored. If an individual's behavior is observed to be rational with respect to a set of rationality conditions, then the behavior is said to be *consistent with rationality (with respect to those conditions)*; if the behavior violates one or more of conditions of the set, then the behavior is said to be *inconsistent with rationality.* It is generally recognized that an individual may employ irrational strategies but show rational behavior. However, it is generally considered that a violation of a rationality condition implies an irrational strategy. In the literature, the $\sqcup$-monotonicity of $\preceq^*$ has been repeatedly taken as a necessary condition for rational behavior. I find this view problematic; that is, I find it problematic to assert that the failure of $\sqcup$-monotonicity is sufficient to establish that a person's behavior is *inconsistent with rationality.*

Let

- $\mathfrak{E}$ be a lattice of scientific events
- $\mathfrak{B}$ be the outer boolean lattice of $\mathfrak{E}$
- $\mathbb{P}$ be a probability function on $\mathfrak{B}$ with $\mathbb{P}(F) = 0$ if and only if $F = \varnothing$
- $\mathfrak{R} = \langle \mathcal{R}, \subseteq, \Cup, \Cap, \ominus, X, \varnothing \rangle$ be the boolean lattice of refutations of $\mathfrak{E}$
- $\mathbb{B}$ be the restriction of $\mathbb{P}$ to $\mathcal{R}$.

Define $\preceq^*$ on $\mathcal{R}$ as follows: For all S and T in $\mathcal{R}$,

$$S \preceq^* T \text{ iff } \mathbb{B}(S) \leq \mathbb{B}(T) .$$

$\Cup$-monotonicity may fail for $\preceq^*$, that is, there may exist A, B, and C in $\mathfrak{R}$ such that $A \Cap C = B \Cap C = \varnothing$, $B \preceq^* A$, but

$$A \Cup C \prec^* B \Cup C .$$

This happens, for example, when

$$A \Cup C = A \cup C \text{ and } B \Cup C = \ominus \ominus (B \cup C) \supset B \cup C,$$

and

$$\mathbb{P}(A \Cup C) < \mathbb{P}(B \Cup C),$$

because then $\mathbb{P}(A \Cup C) = \mathbb{B}(A \Cup C)$, $\mathbb{P}(B \Cup C) = \mathbb{B}(B \Cup C)$, and thus

$$A \Cup C \prec^* B \Cup C,$$

violating $\Cup$-monotonicity. This failure by itself is not inconsistent with rationality, because the probability assignments via $\mathbb{P}$ to the events in the outer boolean lattice $\mathfrak{B}$ of $\mathcal{E}$ satisfy the Kolmogorov axioms and therefore are "consistent with rationality," and thus the assignments via $\mathbb{P}$ to the boolean sublattice $\mathfrak{R}$ of $\mathfrak{B}$ must also be "consistent with rationality."

Narens (2005) gives the following, more detailed account of this idea: Suppose the $\mathbb{B}$ above is person P's probability function. Narens comments,

> Note that in this situation, we have a complete view of the relationship of $\mathfrak{R}$ to $\mathfrak{B}$ without entering into P's mind. Because of this, it might well be argued that an eccentric interpretation is being given to the join operator $\Cup$ in $\mathfrak{R}$—that is, for P, $\Cup$ is not the set-theoretic correlate of "or," (the latter being, of course, $\cup$). However, other examples can be given where this latter argument does not hold without entering into a person's mind.[2]
>
> Consider the case where the Experimenter E presents events to person Q for evaluation of subjective probability. E can check behaviorally that Q is treating the events as if they were events from a boolean algebra of events. For example, E can give (from E's perspective) the events $A \cap (B \cup C)$ and $(A \cap B) \cup (A \cap C)$ and ask Q if they are "identical." Suppose Q passes this test by treating the events "rationally;" that is, treating them from the Experimenter's perspective like they were from a boolean algebra of events. Of course Q could have an idiosyncratic perspective about the events. For example, through odd socialization Q may be what E would consider to be a rabid empiricist who interprets the stimuli different than E: Whereas E considers events to be platonic—either they occur or they do not occur—Q interprets them as empirical and identifies an occurrence of each event A with its double empirical refutation, $\ominus \ominus A$. Suppose for each event A given by E to Q for "probabilistic evaluation," $\mathbb{P}_{\mathcal{R}}(\ominus \ominus A)$ is Q's response. Based on this behavior, should Q's behavior be considered as "inconsistent with rationality," because $\mathbb{P}_{\mathcal{R}}$ is superadditive on $\mathfrak{R}$ ...? I think not. I believe a different version of "consistent with rationality" is called for: Q's behavior should be considered as consistent with rationality

[2] Narens (2005) provides the following footnote: "Of course, a person can enter his own mind and provide an explanation of his behavior. However, experimental psychologists have found that such verbal reports of subjects to be highly suspect and often unreliable as veridical accounts of how the behaviors were produced—see, for example, Nisbet and Wilson (1977). For this and related reasons, behavioral scientists choose to model directly the behavior of subjects rather than their verbal reports of it or a combination of the two."

> as long as we can conceive of an *isomorphic situation of events* in which, *under isomorphism,* another person could give the same probability estimates to event stimuli as Q and still be considered "rational." Under this version, the demonstration of the non-rationality of Q's behavior would consist of showing that no such isomorphism exists.

Let $\mathfrak{B}$, $\mathfrak{E}$, $\mathfrak{R}$, $\mathbb{P}$, and $\mathbb{B}$ be as above. It is often the case that an individual believes that events A and B are independent, $A \perp B$, without computing $\mathbb{P}(A \cap B) = \mathbb{P}(A)\mathbb{P}(B)$. Consistency between $\perp$ and $\mathbb{P}$ requires that for all elements C and D of $\mathfrak{B}$ such that $C \perp D$, $\mathbb{P}(C \cap D) = \mathbb{P}(C)\mathbb{P}(D)$. The following condition, stated for $\mathfrak{B}$ and called $\cup$-*independence,* is usually taken as a rational qualitative axiom about probability with an independence relation $\perp$: For all A, B, and C in $\mathcal{B}$, if $A \perp B$, $A \perp C$, and $B \cap C = \varnothing$, then $A \perp (B \cup C)$. The analogous condition stated for $\mathfrak{R}$ and formulated in terms of $\perp$ and $\Cup$ is called $\Cup$-*independence.* The holding of $\cup$-independence for $\mathfrak{B}$ does not imply the holding of $\Cup$-independence for $\mathfrak{R}$, because there may be A, B, and C in $\mathcal{R}$ such that

$$A \perp B, \quad A \perp C, \quad B \cap C = \varnothing, \quad B \Cup C = \ominus\ominus(B \cup C) \supset B \cup C,$$

and therefore $A \perp (B \cup C)$ by $\cup$-independence but not $A \perp (B \Cup C)$, because

$$\mathbb{P}(A \cap B) + \mathbb{P}(A \cap C) = \mathbb{P}(A) \cdot \mathbb{P}(B \cup C) \neq \mathbb{P}(A) \cdot \mathbb{P}(B \Cup C).$$

The conclusion to be drawn from this is that either (*i*) $\Cup$-independence should be defined with an independence relation that is different from $\perp$, or (*ii*) $\sqcup$-independence is not a valid probabilistic concept for general boolean lattices $\langle L, \leq, \sqcup, \sqcap, u, z\rangle$. I interpret (*i*) as saying that the reasoning involving classical logic used by a rational individual for believing $\ominus\ominus F \perp \ominus\ominus G$ cannot always be reformulated using refutational logic.

In the above, the belief function $\mathbb{B}$ on $\mathfrak{E}$ resulted from restricting the probability function $\mathbb{P}$ on $\mathfrak{B}$ to $\mathfrak{E}$. The following theorem shows a version of the reverse process also holds.

Theorem 9.3 *Suppose* $\mathfrak{E} = \langle \mathcal{E}, \subseteq, \cup, \cap, \ominus, X, \varnothing\rangle$ *is a lattice of scientific events,* $\mathfrak{B} = \langle \mathcal{B}, \subseteq, \cup, \cap, -, X, \varnothing\rangle$ *is the outer boolean lattice of* $\mathfrak{E}$*, and* $\mathbb{B}$ *is a function on* $\mathcal{E}$ *such that it satisfies the following three conditions:*

(i) $\mathbb{B}$ *is into* $[0, 1]$.

(ii) $\mathbb{B}(X) = 1$ *and* $\mathbb{B}(\varnothing) = 0$.

(iii) *For all* F *and* G *in* $\mathcal{E}$*, if* $F \cap G = \varnothing$*, then* $\mathbb{B}(F \cup G) = \mathbb{B}(F) + \mathbb{B}(G)$.

Let $\precsim$ be the binary relation on $\mathcal{E}$ such that for all A and B in $\mathcal{E}$,

$$A \precsim B \text{ iff } \mathbb{B}(A) \leq \mathbb{B}(B).$$

In addition, suppose the following:

(iv) *For all functions $\mathbb{B}'$ on $\mathcal{E}$, if $\mathbb{B}'$ satisfies Conditions* (i) *to* (iii) *above and*

$$A \precsim B \text{ iff } \mathbb{B}'(A) \leq \mathbb{B}'(B),$$

then $\mathbb{B} = \mathbb{B}'$.

Then there exists a weak probability function $\mathbb{P}$ on $\mathfrak{B}$ that is an extension of $\mathbb{B}$.

Proof. Let $\mathfrak{Y} = \langle Y, \mathcal{D}, \precsim \rangle$, where $\mathcal{D}$ is a finite boolean algebra and $\mathcal{D} \subseteq \mathcal{B}$. Then using conditions (*i*) to (*iv*) of the statement of the theorem, it is easy to verify (by modifying the proof of Theorem 4.4) that $\mathfrak{Y}$ satisfies the finite cancellation axioms as formulated in Convention 4.1. By Definition 4.8, Theorem 4.8, and Theorem 4.10, $\langle X, \mathcal{B}, \precsim \rangle$ has a weak probability representation $\mathbb{P}$ (Definition 4.10). By Condition (*iv*) of the statement of theorem, the restriction of $\mathbb{P}$ to $\mathcal{E}$ satisfies Conditions (*i*) to (*iii*) of the theorem, and thus by Condition (*iv*), the restrictions of $\mathbb{P}$ to $\mathcal{E}$ is $\mathbb{B}$; that is, $\mathbb{P}$ is an extension of $\mathbb{B}$. □

Condition (*iv*) of Theorem 9.3 is assumed so that theorems of Chapter 4 can be applied. I believe that it is likely that it could be eliminated by redoing various theorems of Chapter 4 so that they apply more directly to Theorem 9.3. Condition (*iv*) is reasonable for idealizations of empirical situations with rich sets of scientific events.

The philosophical import of Theorem 9.3 is that virtual events can be added to the scientific events in $\mathcal{E}$ to form a boolean lattice of events in such a way that the belief function $\mathbb{B}$ defined on scientific events extends to a probability function $\mathbb{P}$ defined on the combined event space generated by scientific and virtual events. If such an extension did not exist, I believe it would be difficult to justify various scientific practices involving the use of Kolmogorov probability theory.

9.5 An Example Involving American Law

Let

- $\mathfrak{E} = \langle \mathcal{E}, \subseteq, \cup, \cap, \ominus, X, \varnothing \rangle$ be a lattice of scientific events
- $\mathfrak{R} = \langle \mathcal{R}, \subseteq, \Cup, \cap, \ominus, X, \varnothing \rangle$ be the boolean lattice of refutations of $\mathfrak{E}$

- $\mathbb{P}$ be a probability function on $\mathfrak{E}$
- for each nonempty A in $\mathcal{E}$, let

$$v(A) = \frac{\mathbb{P}(\ominus \ominus A)}{\mathbb{P}(A)}$$

- and for each B and C in $\mathfrak{E}$ such that B is nonempty and $C \subseteq B$, let

$$\mathbb{B}(C|B) = \frac{\mathbb{P}(C)v(C)}{\mathbb{P}(B)} = \mathbb{P}(C|B)v(C)\,. \tag{9.3}$$

Narens (2005) comments the following about this situation:

> [Equation 9.3] may be interpreted as an individual's belief in C is his or her subjective probability that given a scientific demonstration of the event B, a scientific refutation can be given to the claim that C is scientifically refutable. Although complicated, such conditional beliefs and their associate degrees of belief make sense. They are especially useful in situations where the demonstration of the double refutation of an event A (i.e., $\ominus \ominus A$) is sufficient to take the same action, or have the same effect, as the event A.

He then gives the following example.

> In American law there is a presumption of innocence; that is, in a criminal proceeding, a lawyer needs only to accomplish the double refutation demonstrating innocence in order to free his or her client; that is, the lawyer needs only to refute the claim that it can be demonstrated that the innocence of the client is refutable. For legal purposes, this has the same effect as the demonstration of innocence. Let p_i be the probability of demonstrating innocence, I, and p the probability of demonstrating the double refutation of innocence, $\ominus\ominus I$. Then $p_i \leq p$, because $I \subseteq \ominus\ominus I$. For various legal decisions and actions, for example, plea bargaining, it is p rather than p_i that is used.

In addition to the above notation at the beginning of this section, let

- $\mathfrak{E} = \langle \mathcal{E}, \subseteq, \cup, \cap, \ominus, X, \varnothing \rangle$ be a lattice of scientific events
- $\mathfrak{B} = \langle \mathcal{B}, \subseteq, \cup, \cap, -, X, \varnothing \rangle$ be the outer boolean algebra of $\mathfrak{E}$

- $\mathfrak{R} = \langle \mathcal{R}, \subseteq, \Cup, \cap, \ominus, X, \varnothing \rangle$ be the boolean lattice of refutations of $\mathfrak{E}$
- $\mathbb{P}$ be a probability function on $\mathfrak{B}$
- and $\mathbb{P}_{\mathcal{R}}$ be the belief function that is the restriction of $\mathbb{P}$ to $\mathcal{R}$.

For the following application, make the additional following assumptions:

- $\mathcal{S}$ be the set of initial events that generates $\mathcal{E}$ (Definitions 9.3 and 9.2).
- Elements of $\mathcal{S}$ are new pieces of evidence. For example, a new witness providing testimony during a trial; the outcome of a particular and ongoing DNA test; etc.
- The only evidence that can be presented are elements of $\mathcal{S}$.
- Evidence are indisputable facts.
- The only propositions that are relevant for probabilistic estimations are double refutations.
- The success (= the occurrence) a double refutation depends on evidence.
- $\mathcal{A}$ is an attorney and $\mathbb{P}$ is $\mathcal{A}$'s probability function on $\mathfrak{B}$.

$\mathbb{P}$ describes $\mathcal{A}$'s current degree of beliefs of events in $\mathfrak{R}$. As new evidence comes in, $\mathcal{A}$ will update his or her event space and probability function. Because the success of a double refutation depends only on evidence, $\mathcal{A}$'s probability function is only updated when new evidence E in $\mathcal{S}$ is presented. In this case the updating of an event D in $\mathcal{E}$ takes the form

$$\mathbb{P}(D|E) = \frac{\mathbb{P}(E \cap D)}{\mathbb{P}(E)}.$$

Let $\ominus\ominus A$ and $\ominus\ominus B$ be in $\mathcal{R}$ and

$$\varnothing \subset \ominus\ominus A \subset \ominus\ominus B \subset X,$$

where, of course, X is the sure event.

Under the above assumptions, the following is a natural question: Is

$$\frac{\mathbb{P}(\ominus\ominus A)}{\mathbb{P}(\ominus\ominus B)}$$

a proper updating of $\mathbb{P}$? The answer is "No." The reason is that for $\ominus\ominus B$ to occur, new evidence F needs to be presented, and by hypothesis, $F \in \mathcal{S}$. The occurrence of $\ominus\ominus B$ depends completely on the F occurring. Thus $F \subseteq \ominus\ominus B$. The usual updating of $\mathbb{P}$ for $\ominus\ominus A$ in this circumstance is given by the formula

$$\frac{\mathbb{P}(F \cap \ominus\ominus A)}{\mathbb{P}(F)} = \mathbb{P}((F \cap \ominus\ominus A)|F)\,.$$

Only rarely is this same as

$$\frac{\mathbb{P}(\ominus\ominus A)}{\mathbb{P}(\ominus\ominus B)}\,.$$

Note that $F \cap \ominus\ominus A$ is not necessarily in $\mathcal{R}$. By assumption, only double refutations are relevant for probabilistic estimations. A consequence of this is that the event space $\mathcal{R}$ of double refutations cannot in general be properly updated to form an updated double refutation event space. This is because from the point of view of $\mathfrak{B}$ (the point of view that is being assumed for doing proper updating), $\mathfrak{R}$ should be updated to $\{C \cap F | C \in \mathcal{R}\}$, which in general contains non-refutations. Although this complicates matters a little, it does not present serious difficulties, because for taking actions only updated degrees of belief are needed, such updated degrees can always be computed for events in $\mathfrak{R}$ by conditionalizing on conjunctions of events from $\mathcal{S}$.

The last two examples establish that conditional belief representations of the form

$$\mathbb{B}(A|B) = \mathbb{P}(A|B)v(A)$$

are in some circumstances a basis for a rational assignment of degrees of belief. The argument for rationality in this and the previous example rests on conceiving a circumstance where for all *relevant* events A and B, the conditional degrees of belief $\mathbb{B}(A|B)$ agree with the rational assignments of conditional probabilities in a structurally isomorphic belief situation. The argument for the existence of appropriate applications of such belief representations rests on the effectiveness and correctness of such representations for boolean lattices of refutations of lattices of scientific events.

Chapter 10

The Psychology of Human Probability Judgments

10.1 Introduction

Section 7.5 discussed descriptions of the theories of human probability judgments of Tversky and Koehler (1994) and Rottenstreich and Tversky (1997). These theories, combined with theoretical extentions and empirical tests, are known in the literature collectively as *Support Theory.* Section 7.5 also presented a model of Narens (2003) for describing Support Theory's empirical phenomena that is based on belief support functions.

This Chapter describes a different approach to Support Theory's phenomena. It is based on changing Support Theory's event space from a boolean algebra of sets to a set of open sets from a topology. The change allows for a new modeling of key aspects of the presumed cognitive processing involved in probabilistic judgments. The Chapter's approach also makes a sharp distinction between the use of semantical representations of descriptions of events as part of natural language processing and the use of cognitive representations of event descriptions in making probability judgments. The semantic representations are modeled as sets in a boolean algebra of sets, while the cognitive representations are modeled as open sets in a topology. The additional structure provided by topology, particularly the relationship between an open set and its boundary, is used to capture important differences in the ways descriptions are represented cognitively.

It is assumed that each description α produces a cognitive representation A that is evaluated in terms of supporting evidence measured by support functions S^+ and S^-. $S^+(A)$ measures the support in favor of A, and $S^-(A)$ measures the support against A. Like in Support Theory, the probability estimate for α is described by the formula,

$$\frac{S^+(A)}{S^+(A) + S^-(A)} . \tag{10.1}$$

But unlike in Support Theory, which models $S^-(A)$ as the support for the boolean complement of A, $-A$, it is assumed that the support against A is modeled as $S^+(\dot{-}A)$, where $\dot{-}A$ is an open set belonging to a topology.

Section 10.2 briefly describes and illustrates the judgmental heuristics of Kahneman and Tversky. These heuristics are part of the foundational basis of Support Theory and appear in many explanations as phenomena generated by judgments of probability. This is followed by a brief summary of the concepts of Support Theory.

Section 10.3 presents the new formulation for Support Theory's phenomena. The new formulations is based on a topology of open sets, and provides an alternative to the theories and explanations in the literature for Support Theory phenomena

Sections 10.4 to 10.6 provide refinements of the new formulation with illustrations from the empirical literature. Section 10.7 provides further considerations about the use of topology in the modeling of human judgments of probability.

10.2 Support Theory

10.2.1 Judgmental Heuristics

Explanations of Support Theory empirical findings generally employ the cognitive heuristics of availability, representativeness, adjustment and anchoring, as described, for example, in Tversky and Kahneman (1974). These heuristics were singled out by Tversky and Koehler (1994) as especially important for their Support Theory. In the new formulation presented in this chapter, only availability and representativeness are employed, with representativeness being reformulated in terms of availability of features.

Tversky and Kahneman (1974) introduce *availability* as follows:

> There are situations in which people assess the frequency of a class or the probability of an event by the ease with which instances or occurrences can be brought to mind. For example, one may assess the risk of heart attack among middle-aged people by recalling occurrences among one's acquaintances. Similarly, one may evaluate the probability that a given business venture will fail by imagining various difficulties it could encounter.

> This judgmental heuristic is called availability. Availability is a useful clue for assessing frequency or probability, because instances of large classes are usually recalled better and faster than instances of less frequent classes. However, availability is affected by factors other than frequency and probability. *(p. 1127)*

They illustrate and described *representativeness* as follows:

> For an illustration of judgment by representativeness, consider an individual who has been described by a former neighbor as follows: "Steve is very shy and withdrawn, invariably helpful, but with little interest in people, or in the world of reality. A meek and tidy soul, he has a need for order and structure, and a passion for detail." How do people assess the probability that Steve is engaged in a particular occupation from a list of possibilities (for example, farmer, salesman, airline pilot, librarian, or physician)? How do people order these occupations from most to least likely? In the representativeness heuristic, the probability that Steve is a librarian, for example, is assessed by the degree to which he is representative of, or similar to, the stereotype of a librarian. Indeed, research with problems of this type has shown that people order the occupations in exactly the same way (Tversky and Kahneman, 1974). This approach to judgment of probability leads to serious errors, because similarity, or representativeness, is not influenced by several factors that should affect judgments of probability. *(p. 1124)*

Tversky and Kahneman describe adjustment and anchoring as follows:

> In many situations, people make estimates by starting from an initial value that is adjusted to yield the final answer. The initial value, or starting point, may be suggested by the formation of the problem, or it may be the result of a partial computation. In either case, adjustments are typically insufficient. That is, different starting points yield different estimates, which are biased toward the initial value. *(p. 1128)*

10.2.2 Empirical Basis

As discussed in Section 7.5, Support Theory was originally designed to explain the following two empirical results, which together appear puzzling:

- *Binary complementarity:* For a binary partition of an event, the sum of the judged probabilities of elements of the partition is 1.
- *Subadditivity:* for partitions consisting of three or more elements, the sum of the judged probabilities of the elements is ≥ 1, with > 1 often being observed.

Section 7.5 presented a study of Fox and Birke (2002) illustrating the simultaneous holding of Binary Complementarity and Subadditivity. There are numerous studies in the literature, run on a variety of subject populations under various experimental conditions, showing the simultaneous holding of Binary Complementarity and Subadditivity. The effects of Subadditivity is often striking, especially when found in professions where one would expect experts to give relatively accurate probability judgments.

Example 10.1 (Medical Diagnosis) Redelmeier, Koehler, Liberman, and Tversky (1995) presented the following scenario to a group of 52 expert physicians at Tel Aviv University.

> B. G. is a 67-year-old retired farmer who presents to the emergency department with chest pain of four hours' duration. The diagnosis is acute myocardial infarction. Physical examination shows no evidence of pulmonar edema, hypothension, or mental status changes. His EKG shows ST-ssegment elevation in the anterior leads, but no dysrythmia or heart block. His past medical history is unremarkable. He is admitted to the hospital and treated in the usual manner. Consider the possible outcomes.

Each physician was randomly assigned to evaluate one the following four prognoses for this patient:

- dying during this admission
- surviving this admission but dying within one year
- living for more than one year but less than ten years
- surviving for more than ten years.

Redelmeier, et al. write,

> The average probabilities assigned to these prognoses were 14%, 26%, 55%, and 69%, repsectively. According to standard theory, the probabilities assigned to these outcomes should sum to 100%. In contrast, the average judgments added to 164% (95% confidence interval: 134% to 194%). □

Another example is an experiment of Fox, Rogers, and Tversky (1996), who asked professional option traders to judge the probability that the closing price of Microsoft stock would fall within a particular interval on a specific future date. They found subadditivity: For example, when four disjoint intervals that spanned the set of possible prices were presented for evaluation, the sums of the assigned probabilities were typically about 1.50. And they also found binary complementarity: When binary partitions were presented the sums of the assigned probabilities were very close to 1, e.g., .98.

Other researchers, e.g., Wallsten, Budescu, and Zwick (1992) have observed Binary Complementarity in laboratory settings.

10.2.3 Summary of the Theory

As discussed in Section 7.5, Support Theory assumes a description is evaluated in terms of a support function s, from a ratio scale family of functions. The support for a description α, $s(\alpha)$, is evaluated through the use of judgmental heuristics. The data for most Support Theory studies consist of judged (conditional) probability of descriptions of the form, "α occurring rather than γ occurring," in symbols, $P(\alpha, \gamma)$, where the logical conjunction of α and γ is understood by the participant to be impossible. It is additionally assumed[1] that

$$P(\alpha, \gamma) = \frac{s(\alpha)}{s(\alpha) + s(\gamma)} .$$

Descriptions of impossible events are called "null." Support Theory also makes a distinction between "implicit" and "explicit" disjunctions, where explicit disjunctions have the form $\alpha \vee \gamma$ ("α or γ") for some nonnull α and γ such that the conjunction of α and γ is null. A description is implicit if and only if it is nonnull and not explicit. Support Theory employs the following concept of "extension:" Two descriptions have the same extension if and only if they describe the same event. Support Theory is founded on the principle that descriptions with the same extension can have different support values assigned to them.

The original version of the Support Theory of Tversky and Koehler (1994) has been generalized in various ways. For the purposes of this chapter, the most important generalization in the literature is that of Rottenstreich and Tversky (1997). The two versions differ as follows: Suppose α

[1] An exception is a study of Idson, Krantz, Osherson, and Bonini, 2001, where a different equation is used to model Support Theory data.

is implicit, $\delta \vee \gamma$ is explicit, and α and $\delta \vee \gamma$ have the same extension. Then the Support Theory of Tversky and Koehler assumed

(1) *implicit subadditivity:* $s(\alpha) \leq s(\delta \vee \gamma)$, and

(2) *explicit additivity:* $s(\delta \vee \gamma) = s(\delta) + s(\gamma)$.

The Support Theory of Rottenstreich and Tversky assumed

(1) *implicit subadditivity:* $s(\alpha) \leq s(\delta \vee \gamma)$, and

(2*) *explicit subadditivity:* $s(\delta \vee \gamma) \leq s(\delta) + s(\gamma)$.

Rottenstreich and Tversky observed cases where explicit additivity (2) failed but explicit subadditivity (2*) held. A direct consequence of (1) and (2*) is

$$s(\alpha) \leq s(\delta) + s(\gamma),$$

which, of course, is weaker than explicit additivity, $s(\delta \vee \gamma) = s(\delta) + s(\gamma)$.

By definition, $\delta \vee \gamma$ is said to be an *unpacking* of α if and only if $\delta \vee \gamma$ is an explicit disjunction that has the same extension as α. α may have many unpackings. The following situation involving unpacking occurs frequently in the Support Theory literature: Nonnull descriptions θ and α are found such that the conjunction of θ and α is impossible and an unpacking $\delta \vee \gamma$ of α is found such that

$$P(\alpha, \theta) < P(\delta \vee \gamma, \theta).$$

In Support Theory, several different kinds of cognitive processes can be used to determine the support for a description:

> The support associated with a given [description] is interpreted as a measure of the strength of evidence in favor of this [description] to the judge. The support may be based on objective data (e.g., frequency of homicide in the relevant data) or on a subjective impression mediated by judgmental heuristics, such as representativeness, availability, or anchoring and adjustment (Kahneman, Solvic, and Tversky, 1982). For example, the hypothesis "Bill is an accountant" may be evaluated by the degree to which Bill's personality matches the stereotype of an accountant, and the prediction "An oil spill along the eastern coast before the end of next year" to be assessed by the ease with which similar accidents come to mind. Support may also reflect reasons or arguments recruited by the judge in favor of

> the hypothesis in question (e.g., if the defendant were guilty, he would not have reported the crime). *(Tversky and Koehler, 1994, p. 549)*

Most of the research of support theorists has focused on the differential effects different kinds of heuristics have on probability judgments.

10.2.4 Focal Descriptions and Contexts in Support Theory

In Support Theory, participants are presented with a description β that establishes the context for the probabilistic judgment of a description α. In this situation, α is called the *focal* description, and consistent with Chapter 7, β is called the *context (description).* Support Theory studies are almost always designed so that the context β contains a description γ, called the *alternative* description, such that it is clear to the participant that the intersection of the extensions of α and γ is empty, that either α or γ must occur, and the extension of β is the same as the extension of $\alpha \vee \gamma$. Throughout this chapter, this situation is described as, *the participant is asked to judge* $\alpha \mid \beta$, *"the probability of* α *given* β*."* In addition, in within-subject designs the participant is asked to judge $\gamma \mid \beta$, and in between-subjects designs other participants are asked to judge $\gamma \mid \beta$.

For a binary description partitions (α, γ), both $\alpha \mid \beta$ and $\gamma \mid \beta$ are presented for evaluation. For a 3-ary description partition (α, δ, γ), all three of $\alpha \mid \beta$, $\delta \mid \beta$, and $\gamma \mid \beta$ are presented, where

$$\text{(the extension of } \alpha\text{, the extension of } \delta\text{, the extension of } \gamma\text{)}$$

is a partition of the extension of $\beta =$ the extension of $(\alpha \vee \delta) \vee \gamma$. Similar assumptions about presentations hold for n-ary description partitions for $n > 3$.

10.3 Concepts Used in the New Formulation

10.3.1 Focal Descriptions and Contexts in the New Formulation

The new formulation employs focal descriptions and contexts much in the same manner as Support Theory. However, there are important differences. The new formulation has two kinds of extensions. One is used for evaluating the support of descriptions. For this extension, the new formulation does not demand that the extension of the context β for probabilistic judgments for a focal description α and its alternative description δ be the same as the extension of their explicit disjunction, $\alpha \vee \delta$. This difference allows for a wider range or probabilistic judgments and modeling possibilities than in Support Theory.

10.3.2 Semantical Representations

The descriptions are assumed to be formulated in a natural language, for example, English. They can be descriptions of individuals, categories, or propositions. However, only descriptions of propositions are presented to the participant for probabilistic evaluation. The natural language is assumed to have the logical operations of conjunction ("and"), disjunction ("or"), and negation ("not"), in symbols, respectively $\wedge$, $\vee$, and $\neg$. Let $\boldsymbol{P}$ be the set of propositions of the natural language. It is assumed that $\boldsymbol{P}$ together with $\wedge$, $\vee$, and $\neg$ is a classical propositional logic. For propositions α and δ in $\boldsymbol{P}$, let $\alpha \sim \delta$ stand for α is logically equivalent to δ. Then $\sim$ is an equivalence relation of $\boldsymbol{P}$. For each α in $\boldsymbol{P}$, let $\alpha^{\sim}$ be the $\sim$-equivalence class to which α belongs. Let $\boldsymbol{P}^{\sim}$ be the set of $\sim$ equivalence classes of $\boldsymbol{P}$. By convention, for A and B in $\boldsymbol{P}^{\sim}$, let $A \wedge B$ be the equivalence class $(\alpha \wedge \delta)^{\sim}$, where α is some element of A and δ is some element of B. Define $A \vee B$ and $\neg A$ in the obvious analogous manners. Let *true* be the equivalence class $(\alpha \vee \neg \alpha)^{\sim}$, where α is some element of $\boldsymbol{P}$, and *false* be the equivalence class $(\alpha \wedge \neg \alpha)^{\sim}$, where α is some element of $\boldsymbol{P}$. Let $\leq$ be the partial ordering on $\boldsymbol{P}$ such that for all A and B in $\boldsymbol{P}^{\sim}$,

$$A \leq B \text{ iff for some } \alpha \text{ in } A \text{ and } \delta \text{ in } B,\ (\alpha \wedge \delta)^{\sim} = \alpha^{\sim}.$$

Then

$$\mathfrak{P} = \langle \boldsymbol{P}, \leq, \vee, \wedge, \neg, true, false \rangle$$

is a boolean lattice. By the Stone Representation Theorem, Theorem 8.4, let φ be an isomorphism of $\mathfrak{P}$ onto a boolean lattice of subsets,

$$\mathfrak{X} = \langle \mathcal{X}, \subseteq, \cup, \cap, -, X, \varnothing \rangle .$$

The natural language semantics provides an abstract representation that associates with each propositional description α, a *semantical representation* of α, $\mathbf{SR}(\alpha)$, which is assumed to be the event $\varphi(\alpha^{\sim})$ of $\mathfrak{X}$. It is assumed that the logical operations of conjunction, disjunction, and negation are interpreted in the natural language representation as intersection, union, and complementation, that is, for all α and β in $\boldsymbol{P}$,

$$\mathbf{SR}(\alpha \wedge \beta) = \varphi((\alpha \wedge \beta)^{\sim}) = \varphi(\alpha^{\sim}) \cap \varphi(\beta^{\sim}) = \mathbf{SR}(\alpha) \cap \mathbf{SR}(\beta),$$

$$\mathbf{SR}(\alpha \vee \beta) = \varphi((\alpha \vee \beta)^{\sim}) = \varphi(\alpha^{\sim}) \cup \varphi(\beta^{\sim}) = \mathbf{SR}(\alpha) \cup \mathbf{SR}(\beta),$$

and

$$\mathbf{SR}(\neg\, \alpha) = \varphi((\neg\, \alpha)^{\sim}) = -\, \varphi(\alpha^{\sim}) = -\, \mathbf{SR}(\alpha).$$

Definition 10.1 The new formulation reformulates some of the central concepts of Support Theory as follows: Let α, δ, γ, and θ be arbitrary descriptions. Then,

- $\mathbf{SR}(\alpha)$ is the *semantical extension* of α;
- α and δ are *semantically disjoint* if and only if $\mathbf{SR}(\alpha) \cap \mathbf{SR}(\delta) = \varnothing$;
- α is *(semantically) null* if and only if $\mathbf{SR}(\alpha) = \varnothing$;
- $\gamma = \alpha \vee \delta$ is an *explicit disjunction* if and only if α and δ are semantically disjoint and nonnull and γ is the description "α or δ".
- $\alpha \vee \delta$ is said to be an *unpacking of* θ if and only if $\alpha \vee \delta$ is an explicit disjunction and $\mathbf{SR}(\alpha \vee \delta) = \mathbf{SR}(\theta)$. □

10.3.3 Cognitive Representations

The natural language semantics plays a relatively minor role in the chapter. Its primary function is to classify propositions. Another form of "representation" plays the major role.

Definition 10.2 When $\alpha \mid \beta$ is presented to the participant for probabilistic judgment, she forms a *cognitive representation* $\mathbf{CR}(\alpha)$ corresponding to α that she uses in judging the probability of $\alpha \mid \beta$.

Cognitive representations are modeled as open sets within a topology $\mathcal{U}$. Unless otherwise specified, the universal set of this topology is an open set that depends on the conditioning event β. This universal set is denoted by Ω_β or just Ω with β understood by context. □

All descriptions employed in computing the probability of $\alpha \mid \beta$ are represented as open subsets of the universal set. The computation, however, may require the creation of additional cognitive representations that, in general, need not correspond to descriptions, and these created representations are also represented as open subsets of Ω.

Definition 10.3 Let α and δ be arbitrary descriptions. Then,

- $\mathbf{CR}(\alpha)$ is the *cognitive extension* of α; and

- α and δ are *cognitively disjoint* if and only if $\mathbf{CR}(\alpha) \cap \mathbf{CR}(\delta) = \varnothing$. □

The following very minimal relationship between the semantic and cognitive representations of a description α is assumed:

$$\mathbf{CR}(\alpha) = \mathbf{SR}(\alpha) \text{ iff } \mathbf{SR}(\alpha) = \varnothing .$$

This assumption—that only the empty set is common to both the semantic and cognitive representations—is possible because of the kind of role the semantics of descriptions plays in the general modeling of judgments of probability considered in the chapter. In fact, in some situations **SR** and **CR** are so unrelated that for some descriptions α and γ,

$$\mathbf{SR}(\alpha) \subset \mathbf{SR}(\gamma) \text{ and } \mathbf{CR}(\gamma) \subset \mathbf{CR}(\alpha) .$$

However, for some kinds of judgments, additional relationships between the semantic and cognitive representations need to hold. As discussed later in Subsection 10.4, this is the case for judgments based on the frequency that exemplars come to mind. For such judgments the following relationship is needed.

Definition 10.4 *Cognitively preserve semantical disjointness* is said to hold for a judgment of probability if and only if for all descriptions σ and τ used in the judgment, if σ and τ are semantically disjoint, then they are cognitively disjoint. □

One of Support Theory's uses of "non-extensionality" is essentially captured in the following definition: Descriptions α and δ are said to exhibit *non-extensionality* if and only if

$$\mathbf{SR}(\alpha) = \mathbf{SR}(\delta) \text{ iff } \mathbf{CR}(\alpha) \neq \mathbf{CR}(\delta) .$$

Note that the above definition "non-extensionality" is based on distinguishing the psychologically based concepts of semantic and cognitive representations.

In the theory of cognitive representations presented in this chapter, the essential features of the cognitive processing in probability judgments are captured by particular cognitive operations performed on cognitive representations and simple relationships among cognitive representations. Open sets from a topology are useful for modeling cognitive representations for a number of reasons. Two of these are (i) in judging probabilities participants need to construct a complement of a cognitive representation, and this better modeled as a kind of open set in a topology than a complement in a boolean algebra of sets; and (*ii*) forms of representational ambiguity and unrealization need to be accounted for, and topological boundaries are a useful device for accomplishing this.

Convention 10.1 Throughout the chapter it is assumed that in evaluating the probability of $\alpha \mid \beta$, "the probability of α occurring given β," the participant, through the use of information presented to her and her own knowledge, creates the *cognitive complement* of $A = \mathbf{CR}(\alpha)$ with respect to the universe under consideration, Ω. It should be noted that in many cases the cognitive complement of A will not correspond to a description. Thus in particular, it is not assumed that the cognitive representation of $\neg A$ is the cognitive complement of A.

It is assumed that the cognitive complement of a description is an open subset of the universe under consideration. By convention, $\dot{-}$ stands for the function that assigns to a cognitive representation of a description under consideration its cognitive complement. $\dot{-}$ is called the *function of cognitive complementation.* By convention, $\cap$ and $\cup$ dominate $\dot{-}$ in abbreviated expressions; for example, $\dot{-}\mathbf{CR}(\alpha) \cap \mathbf{CR}(\alpha)$ is an abbreviated form of $[\dot{-}\mathbf{CR}(\alpha)] \cap \mathbf{CR}(\alpha)$. □

10.3.4 Cognitive Model

Convention 10.2 It is assumed that the participant employs methods of evaluations and heuristics to determine the support for A and the support against A and reports her probability of $\alpha \mid \beta$ as the number,

$$\frac{\text{support for } A}{\text{support for } A \ + \text{support against } A}\,.$$

The New Formulation assumes that the support against A = the support for the cognitive complement of A; that is,

$$\text{support against } A = \text{support for } \dot{-} A.$$

Throughout the chapter, the following notation is used:

- $S^+(A)$ stands for the support for A,
- $S^-(A)$ for the support against A, and
- $\mathbb{P}(\alpha \mid \beta)$ for the participant's judged probability when presented $\alpha \mid \beta$, where $\alpha \neq \varnothing$, $\beta \neq \varnothing$, and $\alpha \subseteq \beta$.

Unless explicitly stated otherwise, the following assumptions are made:

(1) S^+ and S^- are functions into $\mathbb{R}^+$.

(2) $S^-(A) = S^+(\dot{-}A)$.

(3) The participant judges $\mathbb{P}(\alpha \mid \beta)$ in a manner consistent with the equation,

$$\mathbb{P}(\alpha \mid \beta) = \frac{S^+(A)}{S^+(A) + S^+(\dot{-}A)},$$

where $A = \mathbf{CR}(\alpha)$.

(4) For all cognitive representations E and F, if $E \subset F$ and $S^+(E)$ and $S^+(F)$ have been determined, then $S^+(E) < S^+(F)$. □

The Cognitive Model for probability judgments presented in the chapter is divided into two kinds of modeling situations. The first is the modeling of a single probabilistic judgment $\mathbb{P}(\alpha \mid \beta)$. The goal for this situation is to explain the single judgment in terms of judgmental heuristics and support. The second is the modeling of multiple probabilistic judgments with a common conditioning event, for example, the simultaneous modeling of $\mathbb{P}(\alpha \mid \beta)$, $\mathbb{P}(\gamma \mid \beta)$, and $\mathbb{P}(\delta \mid \beta)$, where $\gamma \vee \delta$ is an unpacking of α. The second form of modeling uses inputs from the first, for example, the cognitive representations support functions used in the single modelings. Both kinds of modeling employ in different ways the following important theoretical notion of "clear instance of α."

10.3.5 Clear Instances

Various kinds of instances of the context description are used in determining the support for and against a cognitive representation of a focal description. The most important are clear instances and clear noninstances of a focal description.

Definition 10.5 Let α be a description. Then i is said to be a *clear instance of* α if and only if the following holds: If the participant were to make an independent judgment as to whether i was a "clear exemplar" of α, then the participant would judge i to be a clear exemplar of α.

The set of clear instances of α is denoted by $\mathsf{Cl}(\alpha)$. □

Assumption: *It is assumed throughout this chapter that*

- *the universe under consideration* $= \Omega = \mathsf{Cl}(\beta)$,
- *and* $\mathbf{CR}(\alpha) \subseteq \mathsf{Cl}(\alpha)$. □

Although the concept of "clear instance of α" in Definition 10.5 can be accessed through experiment, most experimental paradigms in probability judgment do not do so. For this reason, "clear instance" in Definition 10.5 is counterfactually defined for such paradigms in terms of a judgment that is not made. In such cases, "clear instance" is a theoretical concept. As discussed below, "clear instance" is directly linked to the availability heuristic.

The following definition provides other notions of "instance."

Definition 10.6 The following are two additional kinds of "instances" of α. The definition assumes a situation where a probability judgment is to be made for $\alpha \mid \beta$.

1. i is said to be a *clear noninstance* of α if and only if i is an element of $\mathsf{CI}(\beta)$ $(= \Omega)$ and the following holds: If the participant were to make an independent judgment as to whether i was not a "clear exemplar" of α, then the participant would judge i to be not a clear exemplar of α. The set of clear noninstances of α is denoted by $\mathsf{CNI}(\alpha)$.

2. i is said to be a *vague instance* of α if and only if $i \in \mathsf{CI}(\beta)$ and i is neither a clear instance of α nor clear noninstance of α. The set of vague instances of α is denoted by $\mathsf{VI}(\alpha)$. □

Note that $\mathsf{CI}(\alpha)$, $\mathsf{CNI}(\alpha)$, and $\mathsf{VI}(\alpha)$ form a partition of the universe Ω $(= \mathsf{CI}(\beta))$. The following definition captures another important kind of clear instance.

Definition 10.7 a is said to be an *unrealized clear instance* of α if and only if

$$a \in \mathbf{Cl}(\alpha),\ a \notin \mathbf{CR}(\alpha) \text{ and } \mathbf{CR}(\alpha) \cup \{a\} \text{ is an open subset of } \Omega\,.$$

The set of unrealized clear instances of $\mathbf{CR}(\alpha)$ is denoted by $\mathsf{UCI}(\alpha)$. □

It follows from Definition 10.7, that each unrealized clear instance i of α either is an isolated boundary point of $\mathbf{CR}(\alpha)$ or is such that $\{i\}$ is an open subset of Ω. It is a result of topology that for each nonempty set T of unrealized clear instances of α, that $\mathbf{CR}(\alpha) \cup T$ is an open subset of Ω. If $\mathbf{CR}(\alpha) \cup T$ is the cognitive representation of a proposition γ, then by a previous assumption about S^+,

$$S^+(\mathbf{CR}(\alpha)) < S^+(\mathbf{CR}(\alpha) \cup T) = S^+(\mathbf{CR}(\gamma))\,.$$

By assumption, cognitive representations, clear instances, and unrealized clear instances are related as follows:

1. $\mathbf{CR}(\alpha) \subseteq \mathsf{Cl}(\alpha) \subseteq \mathsf{Cl}(\beta) = \Omega$.
2. $\dot{-}\mathbf{CR}(\alpha) \subseteq \mathsf{CNI}(\alpha)$.
3. $\dot{-}\mathsf{CNI}(\alpha) \cap \mathbf{CR}(\alpha) = \varnothing$.
4. $\mathsf{UCI}(\alpha) \subseteq \mathsf{Cl}(\alpha)$.

An immediate consequence of items 2 and 3 is that

$$\dot{-}\mathbf{CR}(\alpha) \cap \mathbf{CR}(\alpha) = \varnothing .$$

Let $A = \mathbf{CR}(\alpha)$. Then it is easy to see that $S^+(\dot{-}A)$ depends on the universe under consideration, Ω_β, and therefore on β. Because of this dependence, S^+ should be indexed by the context β as in S^+_β. However, in most cases, the judgments under consideration will have the same context, and thus, by convention, in such cases the subscripting can be omitted.

This is how unrealized instances become important for Support Theory phenomena: Suppose in judging $\mathbb{P}(\alpha \mid \beta)$, the participant has an unrealized clear instance i of α that was not available in the judging of the support for $\mathbf{CR}(\alpha)$, but became available as a clear instance in the judging of the support for $\mathbf{CR}(\gamma \vee \delta)$, where $\gamma \vee \delta$ is an unpacking of α. This mechanism of making new instances available through unpacking is the principal idea behind why unpacking generally increases support.

10.3.6 Single Probability Judgments

When $\alpha \mid \beta$ is presented to the participant for probabilistic judgment, $\Omega = \mathsf{Cl}(\beta)$ is taken as the universal set of a topology. The participant also forms a cognitive representation A corresponding to α that is an open subset of B. She employs cognitive heuristics to find $S^+(A)$ and $S^-(A)$, and produces a judgment of probability, $\mathbb{P}(\alpha \mid \beta)$, consistent with the formula,

$$\mathbb{P}(\alpha \mid \beta) = \frac{S^+(A)}{S^+(A) + S^-(A)} .$$

It is assumed that in finding $S^-(A)$, she creates the cognitive event $\dot{-}A$ and uses,

$$S^-(A) = \text{support for } \dot{-}A = S^+(\dot{-}A) .$$

Convention 10.3 Throughout this subsection, unless explicitly stated otherwise, it is assumed that a case of a single probability judgment $\mathbb{P}(\alpha \mid \beta)$ is being considered and that $A = \mathbf{CR}(\alpha)$, $\Omega = \mathsf{Cl}(\beta)$, and $\dot{-}A$ is the cognitive complement of A with respect to Ω. $\square$

As an illustration, let

- Ω = the cartesian plane,
- A = the open unit disk minus the origin,
- Cl = the open unit disk,
- and $\dot{-}A = \mathsf{CNl} = (\Omega -$ the closed unit disk).

Suppose the topology consists of

$$\Omega,\ \varnothing,\ A,\ A \cup \{\text{origin}\},\ \dot{-}A,\ A \cup \dot{-}A\,.$$

Then

- $\mathsf{UCl} = \{\text{the origin}\}$,
- and Vl = the unit circle.

In classical probability theory, which is based on measure theory, the boundaries of events have measure 0, that is, they have probability 0. In this sense, they can be "ignored" in calculations involving positive probabilities. In the Cognitive Model, the boundaries may contain subsets that contribute positive support. For example, in the just presented illustration, the origin is a boundary point of A. Because the origin is an unrealized clear instance of α, $A \cup \{\text{origin}\}$ is an open set. One would want

$$S^{+}(A \cup \{\text{origin}\}) > S^{+}(A)\,. \tag{10.2}$$

In Equation 10.2, the origin can be viewed as "contributing positive support." In the just presented illustration, vague instances, which compose the unit circle, are also boundary points of A. The Cognitive Model assumes that subsets of vague instances of A are ignored in the computations of $S^{+}(A)$ and $S^{+}(\dot{-}A)$, and thus such subsets have no impact on the number $\mathbb{P}(\alpha \mid \beta)$, which is computed by the equation

$$\mathbb{P}(\alpha \mid \beta) = \frac{S^{+}(A)}{S^{+}(A) + S^{+}(\dot{-}A)}\,.$$

Ignoring vague instances is a plausible cognitive strategy, because the clear instances have more impact on probability estimates than vague ones, and the ignoring of the vague ones greatly reduces the complexity of the calculation of $S^{+}(A)$ and $S^{+}(\dot{-}A)$.

In the just presented illustration, there is only one type of vague instance—a point on the unit circle, which has the distinctive topological characteristic that it is both a boundary point of the cognitive representation of α and its cognitive complement. In other situations, there may be additional types of vague instances that are distinguishable in terms of their topological characteristics. Vague instances are not classified here, because the chapter limits itself to situations where vague instances are ignored in the computation of support.

10.3.7 Multiple Probability Judgments

Many experiments in support theory investigate the influence of unpacking on probability judgments. This is often done by comparing the probability judgment $\mathbb{P}(\alpha \mid \beta)$ with the sum of probability judgments $\mathbb{P}(\alpha \mid \gamma) + \mathbb{P}(\alpha \mid \delta)$, where $\gamma \vee \delta$ is an unpacking of α. In some studies, the probability estimates $\mathbb{P}(\alpha \mid \beta)$ and $\mathbb{P}(\alpha \mid \gamma) + \mathbb{P}(\alpha \mid \delta)$ come from the same participant; in others, the estimates come from different participants. In either case, the estimates for each description is almost always averaged over participants. The goal of the modeling for multiple probabilistic judgments is to provide a qualitative description of the observed relationships among $\mathbb{P}(\alpha \mid \beta)$, $\mathbb{P}(\gamma \mid \beta)$, $\mathbb{P}(\delta \mid \beta)$, and $\mathbb{P}(\alpha \mid \gamma) + \mathbb{P}(\alpha \mid \delta)$ in terms of the cognitive representations α, β, γ, δ, $\gamma \vee \delta$, and their cognitive complements. The main idea is that the unpacking of α into $\gamma \vee \delta$ makes available in the judgings of $\mathbb{P}(\gamma \mid \beta)$, $\mathbb{P}(\delta \mid \beta)$ and $\mathbb{P}(\gamma \vee \delta \mid \beta)$ unrealized clear instances of α that were not available in the judging of $\mathbb{P}(\alpha \mid \beta)$.

10.4 Frequency Based Judgments

This section presents two studies involving unpacking. The first involves two ways of unpacking of an implicit description. Each way demonstrates explicit strict subadditivity. However, the two ways differ on the kind of implicit subadditivity involved: one way exhibiting strict subadditivity, and the other exhibiting additivity.

The second study also involves two ways of unpacking an implicit description. In this case, one way displays additivity, while the other way demonstrates explicit superadditivity. Both studies involve frequency judgments, and their results are explained in terms of the Cognitive Model.

Example 10.2 (Causes of Death I) Rottenstreich and Tversky (1997) conducted the following study involving 165 Standford undergraduate economic students. Two cases, Case 1 and Case 2, were presented for evaluation, with Case 2 taking place a few weeks after Case 1. Both cases consisted of a questionnaire in which the participants were informed the following:

> Each year in the United States, approximately 2 million people (or 1% of the population) die from a variety of causes. In this questionnaire you will be asked to estimate the probability that a randomly selected death is due to one cause rather than another. Obviously, you are not expected to know the exact figures, but everyone has some idea about the prevalence of various causes of death. To give you a feel for the numbers involved, note that 1.5% of deaths each year are attributable to suicide.

In terms of our notation the following were presented for probabilistic judgment:

$$\alpha \mid \beta,\ \alpha_s \mid \beta,\ \alpha_a \mid \beta,\ (\alpha_s \vee \alpha_a) \mid \beta,\ \alpha_d \mid \beta,\ \alpha_n \mid \beta,\ (\alpha_d \vee \alpha_n) \mid \beta,$$

where,

- β is death,
- α is homicide,
- α_s is homicide by a stranger,
- α_a is homicide by an acquaintance,
- $\alpha_s \vee \alpha_a$ is homicide by a stranger or homicide by an acquaintance,
- α_d is daytime homicide,
- α_n is nighttime homicide,
- $\alpha_d \vee \alpha_n$ is homicide during the daytime or homicide during the nighttime.

In both Cases 1 and 2 the participants were randomly divided into three groups of approximately equal size, with each group making the following judgments:

$$\textit{Case 1} = \begin{cases} \text{judge} & \alpha \mid \beta \\ \text{judge} & (\alpha_s \vee \alpha_a) \mid \beta \\ \text{judge both} & \alpha_s \mid \beta \ \text{ and } \ \alpha_a \mid \beta \end{cases}$$

$$Case\ 2 = \begin{cases} \text{judge} & \alpha \mid \beta \\ \text{judge} & (\alpha_d \vee \alpha_n) \mid \beta \\ \text{judge both} & \alpha_d \mid \beta \text{ and } \alpha_n \mid \beta\,. \end{cases}$$

Rottenstreich and Tversky predicted that $\alpha_s \vee \alpha_a$ was "more likely to bring to mind additional possibilities than $\alpha_d \vee \alpha_n$." They reasoned,

> Homicide by an acquaintance suggests domestic violence or a partner's quarrel, whereas homicide by a stranger suggests armed robbery or drive-by shooting. In contrast, daytime homicide and nighttime homicide are less likely to bring to mind disparate acts and hence are more readily repacked as ["homicide"]. Consequently, we expect more implicit subadditivity in Case 1,
>
> $$\text{i.e., } \mathbb{P}(\alpha_s \vee \alpha_a \mid \beta) - \mathbb{P}(\alpha \mid \beta) > \mathbb{P}(\alpha_d \vee \alpha_n \mid \beta) - \mathbb{P}(\alpha \mid \beta)\,,$$
>
> due to enhanced availability, and more explicit subadditivity in Case 2,
>
> $$\begin{aligned} \text{i.e., } & \mathbb{P}(\alpha_d \mid \beta) + \mathbb{P}(\alpha_n \mid \beta) - \mathbb{P}(\alpha_d \vee \alpha_n \mid \beta) \\ & > \mathbb{P}(\alpha_s \mid \beta) + \mathbb{P}(\alpha_a \mid \beta) - \mathbb{P}(\alpha_s \vee \alpha_a \mid \beta)\,, \end{aligned}$$
>
> due to repacking of the explicit disjunction.

Rottenstreich and Tversky found that their predictions held: Letting $\mathbb{P}$ stand for the median probability judgment, they found

$$\begin{array}{lllll} \textit{Case 1:} & \mathbb{P}(\alpha \mid \beta) = .20 & \mathbb{P}(\alpha_s \vee \alpha_a) = .25 & \mathbb{P}(\alpha_s) = .15 & \mathbb{P}(\alpha_a) = .15 \\ \textit{Case 2:} & \mathbb{P}(\alpha \mid \beta) = .20 & \mathbb{P}(\alpha_d \vee \alpha_n) = .20 & \mathbb{P}(\alpha_d) = .10 & \mathbb{P}(\alpha_n) = .21\,. \end{array}$$

□

Rottenstreich and Tversky's prediction that $\alpha_s \vee \alpha_a$ was "more likely to bring to mind additional possibilities than $\alpha_d \vee \alpha_n$," and their subsequent reasoning for the prediction, corresponds to the ideas about unpacking and unrealized clear instances presented in Subsection 10.3.7. However, in order for their argument to work, some additional assumption is needed relating the supports against α and $\alpha_s \vee \alpha_a$. Applying ideas of Subsection 10.3.7 to this situation, it is natural to assume that

$$\text{the support against } \alpha = \text{the support against } \alpha_s \vee \alpha_a. \tag{10.3}$$

Because Subsection 10.3.7 assumes a topology for representing cognitive representations, this assumption can be made while keeping the cognitive representation of β constant. It appears that for this situation, Rottenstreich and Tversky are assuming

$$\text{the support for } \alpha < \text{the support for } \alpha_s \vee \alpha_a.$$

What is unclear is what other assumptions they are making about the (*i*) constancy of β across judgments, and (*ii*) the size relationship between the support against α and the support against $\alpha_s \vee \alpha_a$. □

For probability judgments based on frequency the following assumption is made:

Additional Cognitive Model Assumption *For frequency judgments based on availability, cognitively preserved semantical disjointness (Definition 10.4) holds; that is, for all descriptions σ and τ used in the judgment, if σ and τ are semantically disjoint, then they are cognitively disjoint.* □

Note that the above Cognitive Model assumption is only for frequency judgments; it may fail for other kinds of judgments. In particular, as is discussed later in Subsection 10.5, it often fails for situations when the probability judgment is based on similarity.

Example 10.3 (Causes of Death II) Sloman, Rottenstreich, Wisniewski, Hadjichristidis, and Fox (2004) asked University of Chicago undergraduates to complete the following item:

> Consider all the people that will die in the U.S. next year. Suppose we select one of these people at random. Please estimate the *probability* that this person's death will be attributed to the following causes.

The students were divided into the groups, *packed, typical,* and *weak-atypical,* which judged the following descriptions:

- *packed:* disease
- *typical:* heart disease, cancer, stroke, or any other disease
- *weak-atypical:* pneumonia, diabetes, cirrhosis, or any other disease

The important difference between the typical and weak-atypical descriptions is that the three most common causes of death are explicitly mentioned in the typical description, whereas less common causes of death are explicitly mentioned in the weak-atypical description. Sloman et al. theorized that in judging the packed condition, participants would naturally unpack it into typical exemplars. Because of this, they predicted that the packed condition and the (unpacked) typical condition would yield approximately the same judged probabilities. Thus for this condition they expected unpacking to yield additivity. They also theorized that unpacking the packed condition into atypical exemplars with weak support such as in the weak-typical condition would capture attention away from the more typical exemplars thus yielding lower support and therefore a lower judged probability. Because of this they predicted that the weak-packed condition would yield *superadditivity,* that is, they predicted the sum of the probabilities for the partition will be less than 1. This latter prediction contradicts the basic principle of Support Theory that either additivity or subadditivity should be observed. The data in their study confirmed their predictions: they found additivity for the typical condition and superadditivity for the weak-atypical condition.

Note that in terms of the Cognitive Model, the unpacking of the packed case into the weak-atypical case provides the example of cognitive representations D (for the packed case) and E (for the unpacked case) such that neither $D \subseteq E$ nor $E \subseteq D$. That is, in this multiple probabilistic judgment situation, the packed D has elements that are unrealized clear instances of the unpacked E, and unpacked E has unrealized clear instances of the packed D. □

Sloman et al. also theorized that in the rare cases where a condition is unpacked into atypical exemplars that have stronger support than typical exemplars, one may observed subadditivity.

The above studies of Sloman et al. and Rottenstreich and Tversky involving frequency judgments demonstrates that different methods of unpacking produces different forms of additivity, with additivity, subadditivity, and superadditivity as possible outcomes.

10.5 Judgments Based on Representativeness

Example 10.4 (Linda) Kahneman and Tversky (1982) gave participants the following description β:

> β: Linda is 31 years old, single outspoken and very bright. She majored in philosophy. As a student she was deeply concerned with the issues of discrimination and social justice, and also participated in anti-nuclear demonstrations.

Participants were asked to rank order the following statements by their probability, using 1 for the most probable and 8 for the least probable. The descriptions denoted by γ and α below are the ones that play the important roles in the discussion presented here.

Linda is a teacher in elementary school
Linda works in a bookstore and takes Yoga classes
Linda is active in the feminist movement
Linda is a psychiatric social worker
Linda is a member of the League of Women voters
γ: Linda is a bank teller
Linda is an insurance salesperson
α: Linda is a bank teller and is active in the feminist movement

Over 85% of participants ranked as more probable that Linda was both a bank teller and a feminist (α) than just a bank teller (γ). This is an example of what in the literature is called the *conjunction fallacy*. According to Kahneman and Tversky, the conjunction fallacy is due to representativeness: "bank teller and is active in the feminist movement" is more a "representative" description of Linda than just "bank teller." □

Cognitive representations take many forms. They all have in common that they are sets, but the sets can have different kinds of elements. For representativeness, the elements of the sets are taken to be exemplifying properties. This choice allows for a better modeling of the similarity concept.

For the purposes of this chapter, it is assumed that the representativeness heuristic is employed for making probability judgments about Linda in the above example.[2] Thus the description of Linda, β, makes available to the participant a set of properties, L, that exemplifies people fulfilling that description. Similarly, the predicate "is a bank teller" makes available a set of properties, T, exemplifying bank tellers, and the predicate "is a

[2]There has been some debate about this point, because minor changes in the wording of the description have inverted the relative ranking of α and γ in other data sets. (See Mellers, Hertwig, and Kahneman, 2001, for examples and a discussion.) Such results do not affect the main thrust of this section. Because the new formulation separates natural language semantics from the semantics employed in the probability estimation process, we can have situations where the natural language semantics remains stable on minor rewording but the semantics involved in the probability estimation does not. In particular, a rewording may change the judgment from a judgment about representativeness into a judgment of frequency, thus changing radically the structure of the cognitive representations without changing the natural language semantics in the sense that the two different wordings have the same natural language consequences.

bank teller and is active in the feminist movement" makes available a set of properties, TF, exemplifying people who are bank tellers and are active in the feminist movement. β, α, and γ are assumed to have the following cognitive representations:

- $\mathbf{CR}(\beta) = L$.
- $\mathbf{CR}(\alpha) = L \cap TF$.
- $\mathbf{CR}(\gamma) = L \cap T$.

The "conjunction fallacy" arises because participants give greater support to $L \cap TF$ than $L \cap T$. Theoretically, this occurs because the properties in $L \cap TF$ are more available to the participant than those in $L \cap T$. Also, because T and TF depend on the availability, it is theoretically most likely that for most participants, $TF - T \neq \varnothing$ and $T - TF \neq \varnothing$.

β does not completely characterize a person (real or fictitious); it only gives some characteristics that a person may have. For modeling frequency judgments, the cognitive representation of β should be interpreted as the set D of exemplars d that (cognitively) satisfies β when d is appropriately substituted for "Linda". For modeling similarity judgments, the cognitive representation of β should be interpreted as the set of properties, L. L may be generated in different ways, for example, as the set of properties common to the elements of D, or as the set of properties cognitively derivable from the description β.

The similarity interpretation of proper noun "Linda" is viewed here as the set properties L. The similarity interpretation of the noun phrase "bank teller" also has a set of properties, T, as its cognitive representation. The similarity interpretation of "Linda is a bank teller," γ, is then $L \cap T$. Note how this differs from the semantical representation of γ: In the semantic representation, (*i*) "Linda" is interpreted as an individual, l, not as a set; (*ii*) the predicate, "is a bank teller," is interpreted as a set t (i.e., the set of bank tellers); and (*iii*) the statement, "Linda is a bank teller," is interpreted as the statement $l \in t$. In summary, for similarity judgments involving a propositional description, the cognitive representation relates subject and predicate through set theoretic intersection, while the semantic representation relates them through set theoretic membership.

The primary theoretical difference between frequency and similarity judgments is summarized in the following Cognitive Modeling assumption.

Additional Cognitive Modeling Assumption *Probability judgments involving frequency are based on the* availability of exemplars *that come to* *mind; whereas, probability judgments involving similarity are based on the* availability of exemplifying properties *that come to mind.* □

Cognitively preserved semantical disjointness (Definition 10.4) is assumed to hold for frequency judgments. However, theoretically, it should routinely fail for similarity judgments: For example, the semantical disjointness of $\mathbf{SR}(\zeta)$ and $\mathbf{SR}(\tau)$ for a pair of descriptions τ and ζ does not preclude the cognitive exemplars of τ and ζ from having available properties in common. In the multiple probabilistic representation condition, this kind of consideration leads to the concept of "strong ambiguity."

Definition 10.8 Suppose ζ and τ are semantically disjoint and representativeness is used in the single probabilistic judgments of $\mathbb{P}(\zeta \mid \beta)$ and $\mathbb{P}(\tau \mid \beta)$. In the judgments $\mathbb{P}(\zeta \mid \beta)$ and $\mathbb{P}(\tau \mid \beta)$, let

$$Z = \mathbf{CR}(\zeta) \quad \text{and} \quad T = \mathbf{CR}(\tau) .$$

Consider a multiple probabilistic judgment situation containing the judgments $\mathbb{P}(\zeta \mid \beta)$ and $\mathbb{P}(\tau \mid \beta)$. Then i is said to be *strongly ambiguous with respect to Z and T* if and only if $i \in (Z \cap T)$, that is, if and only if i is a clear instance of Z and a clear instance of T. □

The following hypothetical example illustrates the theoretical impact of strongly ambiguous elements on probability judgments.

Example 10.5 (Strong Ambiguity) A multiple probability judgment situation is produced by having a participant judge at different times "the probability of ζ occurring rather than τ," and "the probability of τ occurring rather than ζ." It is assumed that ζ and τ are semantically disjoint and that the heuristic of representativeness is being used to obtain the probability judgments. It is also assumed that

$$\zeta \vee \tau = \beta ,$$

and the only relevant cognitive representations in the multiple probability judgment situation are

- $Z = \mathbf{CR}(\zeta)$ and $\dot{-}Z$ from the judging of $\mathbb{P}(\zeta \mid \beta)$,
- $T = \mathbf{CR}(\tau)$ and $\dot{-}T$ from the of judging $\mathbb{P}(\tau \mid \beta)$, and
- $\Omega = \mathbf{CR}(\beta)$.

Let

$$\Sigma = \{i \mid i \text{ is strongly ambiguous with respect to } Z \text{ and } T\}.$$

Suppose $\Sigma \neq \varnothing$ and i is an arbitrary element of Σ. Then $i \notin \dot{-}Z$ (because it is a clear noninstance of Z) and similarly $i \notin \dot{-}T$. Thus, in evaluating the support for Z during the judging of $\mathbb{P}(\zeta \mid \zeta \vee \tau)$, all elements of Σ appear as clear instances of Z.[3] Such elements add to the support for Z. However, elements of Σ do not add to the support against Z, because $\Sigma \cap (\dot{-}Z) = \varnothing$. Similarly, in the judging of $\mathbb{P}(\tau \mid \zeta \vee \tau)$, elements of Σ add to the support for T, but do not add to the support against T. Therefore the support for instances in Σ are evaluated twice in computing the sum,

$$\mathbb{P}(\zeta \mid \zeta \vee \tau) + \mathbb{P}(\tau \mid \zeta \vee \tau),$$

while support for other clear instances of Z or of T are evaluated only once, that is, the support for strongly ambiguous elements are "double counted." Furthermore, the "double counting" makes the sum $\mathbb{P}(\zeta \mid \zeta \vee \tau) + \mathbb{P}(\tau \mid \zeta \vee \tau)$ larger than it would have been if the elements of Σ were only "singly counted." □

10.6 Probability Judgments for Binary Partitions

Support Theory was founded on the empirical findings of Binary Complementarity and subadditivity due to unpacking. Example 10.3 shows a violation of the unpacking part of the foundation by providing an empirical example where unpacking yields superadditivity. Later literature (e.g., Brenner and Rottenstreich, 1999; Macchi, Osherson, and Krantz 1999; Idson, Krantz, Osherson, and Bonni, 2001) has provided examples of violations of binary complementarity. An example of a violation of binary complementary is discussed in this section, especially in reference to the Cognitive Model.

Convention 10.4 Throughout this section the following notation and assumptions are observed:

[3]Obviously, if the participant simultaneously recognizes in the judging of $\mathbb{P}(\zeta \mid \zeta \vee \tau)$ that i is a clear instance of ζ and also a clear instance of τ, then she would see that there is a conflict and re-evaluate the status of i, finally deciding that either (i) it is not a clear instance of ζ and disregard it in the computation of $S^+(Z)$, or (ii) decide it is a clear instance ζ and include it in the computation of $S^+(Z)$. However, because the judgings $\mathbb{P}(\zeta \mid \zeta \vee \tau)$ and $\mathbb{P}(\tau \mid \zeta \vee \tau)$ are separated, the participant may not be aware of such a conflict and therefore have no need to re-evaluate the status of i.

- Participants (in either within-subject or between-subjects designs) are presented $\alpha \mid \beta$ and $\gamma \mid \beta$ for probabilistic judgment.
- In the semantics, the extensions of α and γ are disjoint, that is,

$$\mathbf{SR}(\alpha) \cap \mathbf{SR}(\gamma) = \varnothing .$$

- In the semantics, $\alpha \vee \gamma$ exhausts β, that is,

$$\mathbf{SR}(\alpha) \cup \mathbf{SR}(\gamma) = \mathbf{SR}(\beta) .$$

- A, C, and Ω are respectively the cognitive representations of α, γ, and β. □

The following theorem characterizes the sum $\mathbb{P}(\alpha \mid \beta) + \mathbb{P}(\gamma \mid \beta)$ in terms of S^+ and $\dot{-}$:

Theorem 10.1 *The following three statements are true:*

$$\mathbb{P}(\alpha \mid \beta) + \mathbb{P}(\gamma \mid \beta) \textit{ is } \begin{cases} > 1 & \textit{iff } S^+(A)S^+(C) > S^+(\dot{-}A)S^+(\dot{-}C) \\ = 1 & \textit{iff } S^+(A)S^+(C) = S^+(\dot{-}A)S^+(\dot{-}C) \\ < 1 & \textit{iff } S^+(A)S^+(C) < S^+(\dot{-}A)S^+(\dot{-}C) . \end{cases} \tag{10.4}$$

Proof. The theorem easily follows from the equations,

$$\mathbb{P}(\alpha \mid \beta) = \frac{S^+(A)}{S^+(A) + S^+(\dot{-}A)} \text{ and } \mathbb{P}(\gamma \mid \beta) = \frac{S^+(C)}{S^+(C) + S^+(\dot{-}C)} . \quad \square$$

I view the numerous examples of the holding of binary complementarity (additive binary partitions) in the literature as being due to the construction of experimental situations in which it was natural and easy to interpret C as the cognitive complement of A and vice versa; that is, producing experimental situations where

$$C = \dot{-}A \text{ and } A = \dot{-}C . \tag{10.5}$$

For such situations one would expect, by Theorem 10.1, $\mathbb{P}(\alpha \mid \beta) + \mathbb{P}(\gamma \mid \beta) = 1$, that is, one would expect binary complementarity to hold.

The following example of Brenner and Rottenstreich (1999) provides a rationale and means for producing superadditive binary partitions.

Example 10.6 (Best Picture Nominees) Brenner and Rottenstreich (1999) commented the following about the numerous examples in the literature showing Binary Complementarity:

> Several researchers have found compelling evidence for the descriptive validity of binary complementarity ... Interestingly, however, these researchers examined judgments involving only singleton hypotheses ... We suggest there may be a preference toward singletons in the focal position. For example, judging the probability that *Titanic* will win Best Picture rather than *either As Good As it Gets, Good Will Hunting, L. A. Confidential, or The Full Monty* seems quite natural. However, judging the probability that *either As Good As it Gets, Good Will Hunting, L. A. Confidential, or The Full Monty* will win Best Picture rather than *Titanic* seems more unwieldy. Put differently, it seems natural to compare the likelihood of a single possibility to that of a set of alternatives. On the other hand, it seems awkward to compare the likelihood of a set of possibilities to that of a single alternative. As a result, there may be a greater tendency to repack *either As Good As it Gets, Good Will Hunting, L. A. Confidential, or The Full Monty* or more generally any disjunction, when it is in the focal than in the alternative position.

Brenner and Rottenstreich expected to find departures from binary complementarity of the form $P(S, D) + P(D, S) < 1$.

They ran several experiments testing the dual predictions of binary complementarity for singleton-singleton judgments and violations of binary complementarity for singleton-disjunction pairs. They found the following consistent pattern:

> Sums of judgments for complementary hypotheses are close to 1 when the hypotheses are singletons, and are less than 1 when one of the hypotheses is a disjunction. We observed this pattern in judgments of probability and frequency, and for judgments involving both externally and self-generated hypotheses. *(p. 146)* □

Let α and γ be the following descriptions:

- α: Either As Good As it Gets, Good Will Hunting, L. A. Confidential, or The Full Monty will win Best Picture.
- γ: Titanic will win Best Picture.

Assume the similarity heuristic is used in the judgments of $P(\gamma, \alpha)$ and $P(\alpha, \gamma)$, and assume superadditivity is observed. For a propositional description δ, let

- $s_f(\delta)$ be the support for δ in the focal position, and
- $s_a(\delta)$ be the support for δ in the alternate position.

I interpret Rottenstreich's and Brenner's reasoning for observed superadditivity is that

$$s_a(\alpha) > s_f(\alpha) \text{ and } s_a(\gamma) = s_f(\gamma) \tag{10.6}$$

and

$$P(\gamma, \alpha) = \frac{s_f(\gamma)}{s_f(\gamma) + s_a(\alpha)} \text{ and } P(\alpha, \gamma) = \frac{s_f(\alpha)}{s_f(\alpha) + s_a(\gamma)}.$$

I interpret them suggesting that this occurs because α is interpreted differently in the focal position than in the alternative position. Under these assumptions, the conditional hypotheses $\gamma \vee \alpha$ for $P(\gamma, \alpha)$ and $\alpha \vee \gamma$ for $P(\alpha, \gamma)$ would have different interpretations. The Cognitive Model also permits this conclusion. However, the conclusion about superadditivity can also be obtained through the Cognitive Model *without having the interpretations of the conditional hypothesis change as γ and α change from a focal to an alternate position.* This is because the Cognitive Model uses $\dot{-}$ as its complementation operator rather than $-$:

Let

- $\mathbf{CR}(\beta) = \mathbf{CR}(\alpha \vee \gamma) = \mathbf{CR}(\gamma \vee \alpha)$,
- $A = \mathbf{CR}(\alpha)$,
- and $C = \mathbf{CR}(\gamma)$.

Interpret s_a as S^- and s_f as S^+. Then

$$s_a(\alpha) = S^+(\dot{-}A),\ s_f(\alpha) = S^+(A),\ s_a(\gamma) = S^+(\dot{-}C), \text{ and } s_f(\gamma) = S^+(C).$$

Rottenstreich's and Brenner's reasoning about support applied to Equation 10.6 then yields,

$$S^+(\dot{-}A) > S^+(A) \text{ and } S^+(\dot{-}C) = S^+(C),$$

which by Equation 10.4 yields

$$\mathbb{P}(\alpha \mid \beta) + \mathbb{P}(\gamma \mid \beta) < 1.$$

10.7 Topological Considerations

The theory of probability estimation presented in the chapter is applicable to most of the experimental work in the literature self-described as "support theory." It does not apply to all probability estimation tasks, for example, those that present a partition of events and ask participants "to assign probabilities to each of the alternatives so that the probabilities add to 1." And there are some tasks where it gives a wrong answer, requiring a change in the Cognitive Model's method for estimating support.

The central difference of the modeling methods of the chapter and other psychological theories of human judgments of probability in the literature is that the chapter assigns two kinds of representations to propositions: one designed for modeling the semantic aspects of propositions, such as logical connections between propositions, and one for modeling properties of the cognitive interpretations of propositions used in probabilistic judgement. In the support theory literature, propositions are represented in a boolean algebra of sets. In the chapter's new formulation, the semantic aspects of propositions are modeled in terms of a boolean algebra of sets. However, most aspects that enter into the judging of probabilities are not so modeled. Instead they modeled in terms of open sets in a topology, which allow concepts and results from topology to be used as part of the modeling process.

Topology is a well-developed and rich mathematical subject matter. Topological spaces vary enormously with respect to their qualitative features. This chapter's theory makes minimal assumptions about the topology involved in the modeling of cognitive representations. This allows for improvements of the theory by incorporating additional topological features. Such features can either apply to all cognitive representations or just to a restricted class of representations, for example, representations based on representativeness. The chapter's main use of topological modeling consisted in describing the roles of various kinds of clear instances, clear noninstances, and vague instances in the making of probability judgments. In particular, the concept of "unrealized clear instance," which is modeled as a special kind of subset of the boundary of a cognitive representation, plays an important role in describing the impact of unpacking on probability judgments. Other kinds of subsets of boundaries may also be of use for modeling other aspects of probabilistic judgments, although such modeling was not presented in the chapter.

Recall that the operation of $\cap$-complementation (pseudo complementation) $\ominus U$ for an open set U in topology is defined as the largest open set in $\Omega - U$. $\cap$-complementation was not used for forming cognitive complements in the cognitive model. Instead another operation, $\dot{-}$, was used. $\ominus$ does, however, have an interesting interpretation in the new formulation: For the judgment of $\mathbb{P}(\alpha \mid \beta)$, $\ominus\,\mathbf{CR}(\alpha)$ is the set of all clear instances of β that are clear noninstances of α, that is,

$$\ominus\,\mathbf{CR}(\alpha) = \mathsf{NCI}(\alpha)\,. \tag{10.7}$$

With the following additional assumption, $\ominus$ has an interesting interpretation for the New Formulation.

$\ominus$-Description Assumption: For all propositions α and δ,

$$\text{if } \mathbf{SR}(\alpha) = \mathbf{SR}(\delta), \text{ then } \ominus\,\mathbf{CR}(\alpha) = \ominus\,\mathbf{CR}(\delta)\,. \quad \square$$

By the definitions of **SR**, **CR**, CNI, and Equation 10.7, I find the $\ominus$-Description Assumption to be reasonable. For similar reasons, I also find the following related concept to be reasonable:

$$\text{if } \mathbf{SR}(\alpha) \subset \mathbf{SR}(\delta), \text{ then } \ominus\,\mathbf{CR}(\alpha) \supset \ominus\,\mathbf{CR}(\delta)\,.$$

Definition 10.9 Assume the $\ominus$-Description Assumption. Then for each description α, $\ominus\ominus\,\mathbf{CR}(\alpha)$ is called the *Ω-extension* of α. $\square$

The following theorem is immediate from Definition 10.9.

Theorem 10.2 *Assume the $\ominus$-Description Assumption. Then,*

$$\textit{if } \boldsymbol{SR}(\alpha) = \boldsymbol{SR}(\delta), \textit{ then } \ominus\ominus\boldsymbol{CR}(\alpha) = \ominus\ominus\boldsymbol{CR}(\delta)\,.$$

Proof. Immediate from Definition 10.9 and the concept of clear instance. $\square$

As previously discussed, Support Theory's empirical phenomena consists in having a participant judge $\mathbb{P}(\alpha \mid \beta)$ and $\mathbb{P}(\delta \mid \beta)$ where $\mathbf{SR}(\alpha) = \mathbf{SR}(\delta) \subset \mathbf{SR}(\beta)$. By Statement 2 of Theorem 10.2, this is equivalent to having a participant judge $\mathbb{P}(\alpha \mid \beta)$ and $\mathbb{P}(\delta \mid \beta)$ where $\ominus\ominus\mathbf{CR}(\alpha) = \ominus\ominus\mathbf{CR}(\delta)$. $\ominus\ominus\,\mathbf{CR}(\alpha)$ is the set of clear instances of β that are clear instances of α. A primary difference between $\ominus\ominus\,\mathbf{CR}(\alpha)$ and $\mathbf{SR}(\alpha)$ is that $\ominus\ominus\,\mathbf{CR}(\alpha)$ is derived from the participant's mentation and thus can vary from participant to participant, while $\mathbf{SR}(\alpha)$ is part of the semantics of a natural language that is assumed to be shared among participants, and therefore is assumed not to vary from participant to participant.

Chapter 11

Metamathematical Considerations

11.1 Brief Summary

This chapter investigates foundational issues of probability theory involving axiomatization and the Axiom of Choice. In particular, it is shown that the important probabilistic concepts of the uniqueness of weak probability representations and the existence of a probability representation into the real interval $[0, 1]$ cannot be formulated in "first order languages" of logic. The Axiom of Choice is the most controversial axiom in the history of mathematics. Part of this is due to the counter-intuitive results about probability implied by it. The counter-intuitive results and some of the controversy surrounding them are discussed in Section 11.8.

Theorems of Chapters 3 and 4 show that an ultrapower ${}^*\mathfrak{R}$ of the totally ordered field of real numbers $\mathfrak{R}$ preserves many of the algebraic features of $\mathfrak{R}$. These results are generalized and extended in Section 11.4. To characterize in a general way the kinds of features that are preserved, definitions of "first order language" and "model" are provided respectively in Sections 11.2 and 11.3. Using these definitions and constructions involving ultrafilters similar to those of Chapters 3 and 4, it is shown in Section 11.4 that the ultrapower construction and its generalization preserve the truth of statements formulated in a first order language. These are results of a subarea mathematical logic known as "model theory." They are used to derive other model-theoretic theorems in Sections 11.5 and 11.6, which are then applied to foundational issues involving the axiomatization of probabilistic concepts.

11.2 First Order Languages

Definition 11.1 A first order language L is defined as follows:

The *atomic symbols* of L consist of individual constants, variables, the identity symbol, predicates, connectives, quantifiers, and separation symbols.

The *individual constants* of L consist of a list of individual constant symbols, for example,

$$\mathbf{a},\ \mathbf{b},\ \mathbf{c},\ \mathbf{d},\ \mathbf{c}_1,\ldots,\mathbf{c}_m,\ \mathbf{0},\ \mathbf{1}\,.$$

Cases of no individual constant symbols and arbitrarily many (including non-countably many) are allowed.

The *variables* of L are denumerable in number and consist of

$$\mathrm{x},\ \mathrm{y},\ \mathrm{z},\ \mathrm{u},\ \mathrm{v},\ \mathrm{x}_1,\ \mathrm{y}_1,\ \mathrm{z}_1,\ \mathrm{u}_1,\ \mathrm{v}_1,\ \mathrm{x}_2,\ \mathrm{y}_2,\ \mathrm{z}_2,\ \mathrm{u}_2,\ \mathrm{v}_2,\ \ldots\,.$$

The *identity symbol* of L is $=$.

The *predicates* of L consist of symbols of the form $\boldsymbol{P}(,,,\ldots,)$. The symbol $\boldsymbol{P}(,,,\ldots,)$ is said to be an n-placed predicate symbol if and only if it has n spaces separated by $n-1$ commas, where of course, n is some positive integer. Cases of no predicate symbols and arbitrarily many are allowed. The symbols, $\boldsymbol{P}$, $\boldsymbol{Q}$, $\boldsymbol{P_i}$, $i \in \mathbb{I}^+$, etc., are used to denote predicates symbols.

The *connectives* of L are $\neg$ (negation or "not") and $\wedge$ (conjunction or "and").

The *quantifiers* of L are $(\forall\ )$ (the universal quantifier or "for all") and $(\exists\ )$ (the existential quantifier or "there exists").

The *separation symbols* of L are (and).

Expressions of the form $p = q$, where p and q are individual constants or variables of L, are called *identity atomic formulas of* L. *Predicate atomic formulas* of L are obtained by filling the empty spaces of the predicates of L with individual constants or variables of L. For example, $\boldsymbol{S}(\boldsymbol{a},\boldsymbol{b},\boldsymbol{a},\boldsymbol{a})$ and $\boldsymbol{S}(\boldsymbol{d},x,u_3,\boldsymbol{a})$ are predicate atomic formulas of L. The *atomic formulas* of L consist of the identity and predicate atomic formulas of L.

The *well-formed formulas* of L, *wffs,* are obtained by applying the following three rules:

(i) Atomic formulas are wffs.

(ii) If $\boldsymbol{A}$ and $\boldsymbol{B}$ are wffs, then $(\neg\ \boldsymbol{A})$ and $(\boldsymbol{A} \wedge \boldsymbol{B})$ are wffs.

(*iii*) If $\boldsymbol{A}$ is a wff and α is a variable of L (e.g., α is v), then $(\forall\alpha)\boldsymbol{A}$ and $(\exists\alpha)\boldsymbol{A}$ are wffs, provided $\boldsymbol{A}$ does not already contain one of $(\forall\alpha)$ or $(\exists\alpha)$.

Thus, for example, if $\boldsymbol{P}(x)$ and $\boldsymbol{Q}(x)$ are wffs (where $\boldsymbol{P}$ and $\boldsymbol{Q}$ are predicates of L), then

$$((\exists x)\boldsymbol{P}(x) \wedge \boldsymbol{Q}(x))$$

is a wff, but *by the above definition,*

$$(\forall x)((\exists x)\boldsymbol{P}(x) \wedge \boldsymbol{Q}(x))$$

and

$$((\forall x)(\exists x)\boldsymbol{P}(x) \wedge \boldsymbol{Q}(x))$$

are not wffs.

$\boldsymbol{A}$ is said to be a *sentence* of L if and only if $\boldsymbol{A}$ is a wff such that for each variable α that occurs in $\boldsymbol{A}$, each occurrence of α in $\boldsymbol{A}$ is in a wff of the form $(\forall\alpha)\boldsymbol{B}$ or $(\exists\alpha)\boldsymbol{B}$. Thus

$$(\forall x)((\boldsymbol{P}(x) \wedge \boldsymbol{Q}(y))$$

is not a sentence of L, but

$$(\forall x)(\forall y)((\boldsymbol{P}(x) \wedge \boldsymbol{Q}(y))$$

and

$$(\forall x)((\boldsymbol{P}(x) \wedge (\forall y)\boldsymbol{Q}(y))$$

are sentences of L. □

For readability, square brackets [and] will often be substituted for (and), and sometimes parentheses will be omitted when a formula's correct formation is obvious from context.

Although L has $\neg$ and $\wedge$ as its only connectives, the other familiar logical connectives are definable in terms of them, for example,

$\vee$ ("or") by: $(\boldsymbol{A} \vee \boldsymbol{B})$ if and only if $\neg\,(\neg\boldsymbol{A} \ \wedge\ \neg\boldsymbol{B})$,

$\rightarrow$ ("implies") by: $(\boldsymbol{A} \rightarrow \boldsymbol{B})$ if and only if $(\neg\boldsymbol{A} \ \vee\ \boldsymbol{B})$, and

$\leftrightarrow$ ("if and only if") by:

$$(\boldsymbol{A} \leftrightarrow \boldsymbol{B}) \text{ if and only if } ((\boldsymbol{A} \rightarrow \boldsymbol{B}) \ \wedge\ (\boldsymbol{B} \rightarrow \boldsymbol{A})).$$

Definition 11.2 A first order language L is said to be *countable* if and only if it has countably many individual constant symbols and countably many predicates. □

11.3 Models

Definition 11.3 $\langle Z, R_j, e_k \rangle_{j \in J, k \in K}$ is said to be a *relational structure* if and only if Z is a nonempty set, for each j in J, R_j is some n_j-ary relation on Z, where $n_j \in \mathbb{I}^+$, and for each k in K, e_k is an element of Z. (It is allowed for either J or K to be empty.) □

Definition 11.4 Let L be the first order language with n_j-ary predicates $\boldsymbol{P}_j$ for each j in J, and individual constants $\boldsymbol{c}_k$ for each k in K. $\mathfrak{M}$ is said to be a *model of* L if and only if $\mathfrak{M}$ is an ordered pair $\langle Z, F \rangle$, where

(*i*) Z is a nonempty set, and

(*ii*) F is a function on $\{\boldsymbol{P}_j\}_{j \in J} \cup \{\boldsymbol{c}_k\}_{k \in K}$ such that for each j in J, $F(\boldsymbol{P}_j)$ is an n_j-ary relation on Z, and for each k in K, $F(\boldsymbol{c}_k)$ is an element of Z.

Let $\mathfrak{M} = \langle Z, F \rangle$ be model of L. Then Z is called the *domain (of discourse)* of $\mathfrak{M}$, for each j in J, $F(\boldsymbol{P}_j)$ is called the *interpretation of* $\boldsymbol{P}_j$ in $\mathfrak{M}$, and for each k in K, $F(\boldsymbol{c}_k)$ is called the *interpretation of* $\boldsymbol{c}_k$ in $\mathfrak{M}$. By definition, the *cardinality* of $\mathfrak{M}$ is the cardinality of Z. □

Note that by Definitions 11.3 and 11.4, that if $\langle Z, F \rangle$ is a model of L, then $\langle Z, F(\boldsymbol{P}_j), F(\boldsymbol{c}_k) \rangle_{j \in J, k \in K}$ is a relational structure.

Convention 11.1 Throughout the remainder of this section, let L be the first order language with n_j-ary predicates $\boldsymbol{P}_j$ for j in J, individual constants $\boldsymbol{c}_k$ for k in K, and $\mathfrak{M} = \langle Z, F \rangle$ be a model of L. □

Definition 11.5 A sentence $\boldsymbol{\theta}$ of L is said to be *true* about the model $\mathfrak{M}$, in symbols, $\mathfrak{M} \models \boldsymbol{\theta}$, if and only if the following holds: if

- $=$ is interpreted as the identity relation on Z (i.e., as $=$),
- each predicate $\boldsymbol{P}$ of L occurring in $\boldsymbol{\theta}$ is interpreted as $F(\boldsymbol{P})$,
- the quantifiers, connectives, and separation symbols occurring in $\boldsymbol{\theta}$ are given their usual interpretations (e.g., $\wedge$ is interpreted as "and", etc.),

then the resulting interpreted sentence is a true proposition about the relational structure,

$$\langle Z, F(\boldsymbol{P}_j), F(\boldsymbol{c}_k) \rangle_{j \in J, k \in K} \, .$$

Notice that for sentences $\boldsymbol{\sigma}$ and $\boldsymbol{\tau}$ of L, it follows by the meanings of $\neg$ and $\wedge$ that

$$\mathfrak{M} \models \neg\,\boldsymbol{\sigma} \quad \text{iff} \quad \text{not } \mathfrak{M} \models \boldsymbol{\sigma}\,,$$

and

$$\mathfrak{M} \models \boldsymbol{\sigma} \wedge \boldsymbol{\tau} \quad \text{iff} \quad \mathfrak{M} \models \boldsymbol{\sigma} \text{ and } \mathfrak{M} \models \boldsymbol{\tau}\,.$$

Two sentences $\boldsymbol{\sigma}$ and $\boldsymbol{\tau}$ are said to be *(first order) equivalent*, in symbols, $\boldsymbol{\sigma}$ eq $\boldsymbol{\tau}$, if and only if for each model $\mathfrak{N}$ of L,

$$\mathfrak{N} \models \boldsymbol{\sigma} \quad \text{iff} \quad \mathfrak{N} \models \boldsymbol{\tau}\,. \quad \square$$

It is easy to see that for each sentence $\boldsymbol{\theta}$ of L there exists a sentence $\boldsymbol{\theta}'$ of L such that $\boldsymbol{\theta}$ eq $\boldsymbol{\theta}'$ and such that each variable occurring in $\boldsymbol{\theta}'$ occurs with exactly one occurrence of a quantifier; for example, if $\boldsymbol{\theta}$ is

$$[(\forall x)\boldsymbol{P}(x,\boldsymbol{a}) \,\wedge\, (\exists x)(\forall y)\boldsymbol{Q}(x,y)]\,,$$

where $\boldsymbol{P}$ and $\boldsymbol{Q}$ are predicates of L, then

$$[(\forall x)\boldsymbol{P}(x,\boldsymbol{a}) \,\wedge\, (\exists z)(\forall y)\boldsymbol{Q}(z,y)]$$

is a suitable candidate for $\boldsymbol{\theta}'$.

Definition 11.6 A sentence $\boldsymbol{\theta}$ is said to be in *prenex normal form* if and only if $\boldsymbol{\theta}$ is of the form,

$$(\mathsf{Q}_1 u_1)\cdots(\mathsf{Q}_n u_n)(\boldsymbol{\tau})\,,$$

where $\mathsf{Q}_1,\ldots,\mathsf{Q}_n$ are quantifiers of L, $u_1,\ldots,u_n$ are distinct variables of L, and $\boldsymbol{\tau}$ is a wff of L that does not contain any quantifiers.

Suppose $\boldsymbol{\theta} = (\mathsf{Q}_1 u_1)\cdots(\mathsf{Q}_n u_n)(\boldsymbol{\tau})$ is in prenex normal form. Then $(\mathsf{Q}_1 u_1)\cdots(\mathsf{Q}_n u_n)$ is called the *prefix* of $\boldsymbol{\theta}$, and $\boldsymbol{\tau}$ is called the *matrix* of $\boldsymbol{\theta}$. $\square$

Theorem 11.1 *Each sentence $\boldsymbol{\theta}$ of L is first order equivalent to a sentence $\boldsymbol{\theta}''$ of L that is in prenex normal form.*

Informal proof. First note that if $\boldsymbol{\sigma}$, $\boldsymbol{\tau}$, and $\boldsymbol{\delta}$ are wffs of L such that the variable x does not occur in $\boldsymbol{\delta}$, then the following five statements are true:

(*i*) $\neg\,(\forall x)\boldsymbol{\sigma}$ eq $(\exists x)\neg\,\boldsymbol{\sigma}$.

(*ii*) $\neg\,(\exists x)\boldsymbol{\sigma}$ eq $(\forall x)\neg\,\boldsymbol{\sigma}$.

(iii) $(\boldsymbol{\delta} \wedge (\forall x)\boldsymbol{\sigma})$ eq $(\forall x)(\boldsymbol{\delta} \wedge \boldsymbol{\sigma})$.

(iv) $(\boldsymbol{\delta} \wedge (\exists x)\boldsymbol{\sigma})$ eq $(\exists x)(\boldsymbol{\delta} \wedge \boldsymbol{\sigma})$.

(v) $(\boldsymbol{\sigma} \wedge \boldsymbol{\tau})$ eq $(\boldsymbol{\tau} \wedge \boldsymbol{\sigma})$.

Let $\boldsymbol{\theta}$ be an arbitrary sentence of L. Then a sentence $\boldsymbol{\theta}'$ of L can be found such that $\boldsymbol{\theta}$ eq $\boldsymbol{\theta}'$ and each variable that occurs in $\boldsymbol{\theta}'$ occurs with exactly one occurrence of a quantifier. By the appropriate application of *(i)* to *(v)* above, the quantifiers in $\boldsymbol{\theta}'$ "can be brought to the outside," yielding a sentence $\boldsymbol{\theta}''$ of L that is in prenex normal form and is first order equivalent to $\boldsymbol{\theta}'$, and therefore first order equivalent to $\boldsymbol{\theta}$. The following example illustrates this procedure. Consider the sentence,

$$[(\forall x)(\exists y)\boldsymbol{P}(x,y) \wedge \neg(\exists x)(\exists y)\boldsymbol{Q}(x,y)] .$$

This sentence is first order equivalent to

$$[(\forall x)(\exists y)\boldsymbol{P}(x,y) \wedge \neg(\exists u)(\exists v)\boldsymbol{Q}(u,v)] ,$$

which, by two applications of *(ii)*, is first order equivalent to

$$[(\forall x)(\exists y)\boldsymbol{P}(x,y) \wedge (\forall u)(\forall v)\neg\boldsymbol{Q}(u,v)] ,$$

which, by applications of *(iii)*, *(iv)*, and *(v)*, is first order equivalent to

$$[(\forall x)(\exists y)(\forall u)(\forall v)[\boldsymbol{P}(x,y) \wedge \neg\boldsymbol{Q}(u,v)]] . \quad \square$$

Definition 11.7 Let $\boldsymbol{\theta}$ be a wff of L that does not contain a quantifier. The notation

$$\boldsymbol{\theta}[v_1, \ldots, v_n, \boldsymbol{c}_1, \ldots, \boldsymbol{c}_m]$$

is used to say that the variables that occur in $\boldsymbol{\theta}$ are *among* the variables $v_1, \ldots, v_n$ of L, and that the individual constants that occur in $\boldsymbol{\theta}$ are *among* the individual constants $\boldsymbol{c}_1, \ldots, \boldsymbol{c}_m$. $\square$

Definition 11.8 Suppose $\mathfrak{M} = \langle Z, F \rangle$ is a model of L,

$$\boldsymbol{\theta} = \boldsymbol{\theta}[v_1, \ldots, v_n, \boldsymbol{c}_1, \ldots, \boldsymbol{c}_m]$$

is a formula of L that does not contain a quantifier, and $a_1, \ldots, a_m$ are elements of Z. Then

$$\boldsymbol{\theta}[a_1, \ldots, a_n, \boldsymbol{c}_1, \ldots, \boldsymbol{c}_m]$$

is, by definition, the expression that results from $\boldsymbol{\theta}$ by substituting a_1 for each occurrence of v_1 in $\boldsymbol{\theta}$, ..., by substituting a_n for each occurrence of v_n in θ.

For each $a_1, \ldots, a_n$ in Z,

$$\mathfrak{M} \models \boldsymbol{\theta}[a_1, \ldots, a_n, \boldsymbol{c}_1, \ldots, \boldsymbol{c}_m]$$

is defined inductively as follows: (For purposes of readability, the definition is given in terms of cases for special atomic formulas in (*i*) and (*ii*) below. The case for general atomic formulas is an obvious generalization, and its formulation is left to the reader.)

(*i*) Suppose $\boldsymbol{\theta}$ is an atomic formula involving a predicate symbol, say

$$\boldsymbol{P}(v_3, \boldsymbol{c}_3, \boldsymbol{c}_1, v_1)\,.$$

Let $F(\boldsymbol{P}) = R$, $F(\boldsymbol{c}_1) = e_1$, and $F(\boldsymbol{c}_3) = e_3$. Then, by definition,

$$\mathfrak{M} \models \boldsymbol{\theta}[a_1, \ldots, a_n, \boldsymbol{c}_1, \ldots, \boldsymbol{c}_m] \text{ iff } R(a_3, e_3, e_1, a_1)\,.$$

(*ii*) Suppose $\boldsymbol{\theta}$ is an atomic formula involving the identity symbol, say

$$v_2 = \boldsymbol{c}_3\,.$$

Let $F(\boldsymbol{c}_3) = e_3$. Then, by definition,

$$\mathfrak{M} \models \boldsymbol{\theta}[a_1, \ldots, a_n, \boldsymbol{c}_1, \ldots, \boldsymbol{c}_m] \text{ iff } a_2 = e_3\,.$$

(*iii*) If $\boldsymbol{\theta}$ is $\neg\,\boldsymbol{\sigma}$, then

$$\mathfrak{M} \models \boldsymbol{\theta}[a_1, \ldots, a_n, \boldsymbol{c}_1, \ldots, \boldsymbol{c}_m] \text{ iff not } \mathfrak{M} \models \boldsymbol{\sigma}[a_1, \ldots, a_n, \boldsymbol{c}_1, \ldots, \boldsymbol{c}_m]\,.$$

(*iv*) If $\boldsymbol{\theta}$ is $\boldsymbol{\sigma} \wedge \boldsymbol{\tau}$, then

$$\mathfrak{M} \models \boldsymbol{\theta}[a_1, \ldots, a_n, \boldsymbol{c}_1, \ldots, \boldsymbol{c}_m] \text{ iff }$$
$$\mathfrak{M} \models \boldsymbol{\sigma}[a_1, \ldots, a_n, \boldsymbol{c}_1, \ldots, \boldsymbol{c}_m] \text{ and } \mathfrak{M} \models \boldsymbol{\tau}[a_1, \ldots, a_n, \boldsymbol{c}_1, \ldots, \boldsymbol{c}_m]\,.$$ □

Definition 11.9 (Skolem functions) Suppose $\boldsymbol{P}$ is a 4-place predicate of L and $\boldsymbol{\theta}$ is the sentence

$$(\forall x)(\exists y)(\forall u)(\exists v)\boldsymbol{P}(x, y, u, v)$$

of L. Let $R = F(\boldsymbol{P})$. Then

> $\mathfrak{M} \models \boldsymbol{\theta}$ if and only if for each x in Z there exists y in Z (pick one such y and call it $\varphi(x)$) such that for each u in Z there exists v in Z such that $R(x, y, u, v)$.

Thus $\mathfrak{M} \models \boldsymbol{\theta}$ if and only if there exists a function φ from Z into Z such that for each x and u in Z, there exists v in Z such that

$$R(x, \varphi(x), u, v),$$

that is,

$$\mathfrak{M} \models \boldsymbol{\theta} \text{ iff } (\forall x)(\forall u)(\exists v)\boldsymbol{P}(x, \varphi(x), u, v).$$

Thus the quantifier $(\exists y)$ has been eliminated by the addition of the function φ. Similarly, the quantifier $(\exists v)$ can be eliminated by noting that $\mathfrak{M} \models \boldsymbol{\theta}$ if and only if there exist functions φ from Z into Z and η from $Z \times Z$ into Z such that for each x and u in Z,

$$R(x, \varphi(x), u, \eta(x, u)). \tag{11.1}$$

φ and η in Equation 11.1 are called *Skolem functions.* Their existence follows by the Axiom of Choice (Definition 1.7). The above shows that $\mathfrak{M} \models \boldsymbol{\theta}$ *if and only if there exist Skolem functions* φ *and* η *such that for all* a *and* b *in* Z, $\mathfrak{M} \models \boldsymbol{P}[a, \varphi(a), b, \eta(a, b)]$.

It is a straightforward matter to generalize this result to arbitrary sentences of L that are in prenex normal form. For example, if $\boldsymbol{\sigma}$ is the prenex normal form sentence,

$$(\exists x)(\forall y)(\forall u)(\exists v)\boldsymbol{\tau}(x, y, u, v, \boldsymbol{d})$$

then $\mathfrak{M} \models \boldsymbol{\sigma}$ if and only if there exist an element a of Z and a Skolem function ξ such that for all b and c in Z,

$$\mathfrak{M} \models \boldsymbol{\tau}(a, b, c, \xi(b, c), \boldsymbol{d}). \quad \square$$

11.4 Łoś's Theorem

Definition 11.10 Suppose H, J, and K are sets such that $H \neq \varnothing$ and for each i in H,

$$\mathfrak{X}_i = \langle X^i, R^i_j, e^i_k \rangle_{j \in J,\, k \in K}$$

is a relational structure such that for each j in J, R^i_j is a n_j-ary relation on X^i, and for each k in K, e^i_k is an element of X^i. Let $\mathcal{U}$ be an ultrafilter on H and $\mathcal{X}$ be the set of all functions f on H such that for each i in H, $f(i) \in X^i$.

Define $\sim$ on $\mathcal{X}$ as follows: for all f and g in $\mathcal{X}$,

$$f \sim g \text{ iff } \{i \mid f(i) = g(i)\} \in \mathcal{U}.$$

Then $\sim$ is an equivalence relation on $\mathcal{X}$. Let X be the set of $\sim$-equivalence classes of $\mathcal{X}$.

For each j in J, let R_j be the n_j-ary relation on X defined by:

For each $\alpha_1, \ldots, \alpha_{n_j}$ in X, $R_j(\alpha_1, \ldots, \alpha_{n_j})$ if and only if for some $f_1 \in \alpha_1, \ldots, f_{n_j} \in \alpha_{n_j}$

$$\{i \mid R^i_j(f_1(i), \ldots, f_{n_j}(i))\} \in \mathcal{U}.$$

(Notice that for each j in J, if $f_1 \sim f'_1, \ldots, f_{n_j} \sim f'_{n_j}$, then

$$\{i \mid R^i_j(f_1(i), \ldots, f_{n_j}(i))\} \in \mathcal{U} \text{ iff } \{i \mid R^i_j(f'_1(i), \ldots, f'_{n_j}(i))\} \in \mathcal{U}.)$$

For each k in K, let e_k be the element of X such that $f \in e_k$, where for each i in H, $f(i) = e^i_k$.

Then $\langle X, R_j, e_k \rangle_{j \in J, k \in K}$ is called the *$\mathcal{U}$-ultraproduct of* $\{\mathfrak{X}_i\}_{i \in H}$. □

Definition 11.11 (Ultraproduct) Suppose L is a first order language with predicates $\{P_j\}_{j \in J}$ and individual constants $\{c_k\}_{k \in K}$, H is a nonempty set, and $\mathfrak{M}_i = \langle X_i, F_i \rangle$ is a model of L for each i in H. Then $\langle X, F \rangle$ is said to be the *$\mathcal{U}$-ultraproduct of* $\{\mathfrak{M}_i\}_{i \in H}$ if and only if

$$\langle X, F(\boldsymbol{P}_j), F(\mathbf{c}_k) \rangle_{j \in J, k \in K}$$

is the $\mathcal{U}$-ultraproduct of

$$\{\langle X_i, F_i(\boldsymbol{P}_j), F_i(\mathbf{c}_k) \rangle_{j \in J, k \in K}\}_{i \in H}.$$

It is immediate that $\langle X, F \rangle$ is a model of L. □

Lemma 11.1 *Suppose*

- L *is a first order language*
- $\boldsymbol{\theta}[x_1, \ldots, x_n, \mathbf{c}_1, \ldots, \mathbf{c}_n]$ *is a formula of* L *that contains no quantifiers*
- H *is a nonempty set*
- $\mathcal{U}$ *is an ultrafilter on* H
- *for each* i *in* H, $\mathfrak{M}_i = \langle Z_i, F_i \rangle$ *is a model of* L
- *and* $\mathfrak{M} = \langle Z, F \rangle$ *is the* $\mathcal{U}$*-ultraproduct of* $\{\mathfrak{M}_i\}_{i \in H}$.

Then for each $\alpha_1, \ldots, \alpha_n$ *in* Z *and each* $f_1 \in \alpha_1, \ldots, f_n \in \alpha_n$,

$$\begin{aligned} \mathfrak{M} &\models \boldsymbol{\theta}[\alpha_1, \ldots, \alpha_n, \mathbf{c}_1, \ldots, \mathbf{c}_n] \\ &\textit{iff } \{i \mid \mathfrak{M}_i \models \boldsymbol{\theta}[f_1(i), \ldots, f_n(i), \mathbf{c}_1, \ldots, \mathbf{c}_n\} \in \mathcal{U}. \end{aligned}$$

Proof. (i) Suppose $\boldsymbol{\theta}$ is an atomic formula of L. Then the lemma follows by the definition of "ultraproduct" (Definitions 11.10 and 11.11).

(ii) Suppose the lemma is true for $\boldsymbol{\sigma}[x_1, \ldots, x_n, \mathbf{c}_1, \ldots, \mathbf{c}_n]$ and $\boldsymbol{\theta}$ is $\neg\boldsymbol{\sigma}$. Then

$$\begin{aligned} &\{i \mid \mathfrak{M}_i \models \boldsymbol{\theta}[f_1(i), \ldots, f_n(i), \mathbf{c}_1, \ldots, \mathbf{c}_n]\} \in \mathcal{U} \\ &\text{iff} \quad \{i \mid \mathfrak{M}_i \models \neg\, \boldsymbol{\sigma}[f_1(i), \ldots, f_n(i), \mathbf{c}_1, \ldots, \mathbf{c}_n]\} \in \mathcal{U} \\ &\text{iff} \quad \{i \mid \text{not } \mathfrak{M}_i \models \boldsymbol{\sigma}[f_1(i), \ldots, f_n(i), \mathbf{c}_1, \ldots, \mathbf{c}_n]\} \in \mathcal{U} \\ &\text{iff} \quad \{i \mid \mathfrak{M}_i \models \boldsymbol{\sigma}[f_1(i), \ldots, f_n(i), \mathbf{c}_1, \ldots, \mathbf{c}_n\} \notin \mathcal{U} \\ &\text{iff} \quad \text{not } \mathfrak{M} \models \boldsymbol{\sigma}[\alpha_1(i), \ldots, \alpha_n(i), \mathbf{c}_1, \ldots, \mathbf{c}_n] \\ &\text{iff} \quad \mathfrak{M} \models \neg\, \boldsymbol{\sigma}[\alpha_1(i), \ldots, \alpha_n(i), \mathbf{c}_1, \ldots, \mathbf{c}_n] \\ &\text{iff} \quad \mathfrak{M} \models \boldsymbol{\theta}[\alpha_1(i), \ldots, \alpha_n(i), \mathbf{c}_1, \ldots, \mathbf{c}_n]. \end{aligned}$$

(iii) Suppose the lemma is true for

$$\boldsymbol{\sigma}[x_1, \ldots, x_n, \mathbf{c}_1, \ldots, \mathbf{c}_n] \text{ and } \boldsymbol{\tau}[x_1, \ldots, x_n, \mathbf{c}_1, \ldots, \mathbf{c}_n]$$

and $\boldsymbol{\theta}$ is $\boldsymbol{\sigma} \wedge \boldsymbol{\tau}$. Then

$$\begin{aligned} &\{i \mid \mathfrak{M}_i \models \boldsymbol{\theta}[f_1(i), \ldots, f_n(i), \mathbf{c}_1, \ldots, \mathbf{c}_n]\} \in \mathcal{U} \\ &\quad \text{iff} \quad \{i \mid \mathfrak{M}_i \models \boldsymbol{\sigma}[f_1(i), \ldots, f_n(i), \mathbf{c}_1, \ldots, \mathbf{c}_n] \\ &\qquad\qquad \wedge\, \boldsymbol{\tau}[f_1(i), \ldots, f_n(i), \mathbf{c}_1, \ldots, \mathbf{c}_n]\} \in \mathcal{U} \\ &\quad \text{iff} \quad \{i \mid \mathfrak{M}_i \models \boldsymbol{\sigma}[f_1(i), \ldots, f_n(i), \mathbf{c}_1, \ldots, \mathbf{c}_n] \\ &\qquad\qquad \text{and } \mathfrak{M}_i \models \boldsymbol{\tau}[f_1(i), \ldots, f_n(i), \mathbf{c}_1, \ldots, \mathbf{c}_n]\} \in \mathcal{U} \\ &\quad \text{iff} \quad \{i \mid \mathfrak{M}_i \models \boldsymbol{\sigma}[f_1(i), \ldots, f_n(i), \mathbf{c}_1, \ldots, \mathbf{c}_n]\} \\ &\qquad\qquad \cap \{i \mid \mathfrak{M}_i \models \boldsymbol{\tau}[f_1(i), \ldots, f_n(i), \mathbf{c}_1, \ldots, \mathbf{c}_n]\} \in \mathcal{U} \\ &\quad \text{iff} \quad \{i \mid \mathfrak{M}_i \models \boldsymbol{\sigma}[f_1(i), \ldots, f_n(i), \mathbf{c}_1, \ldots, \mathbf{c}_n]\} \in \mathcal{U} \\ &\qquad\qquad \text{and } \{i \mid \mathfrak{M}_i \models \boldsymbol{\tau}[f_1(i), \ldots, f_n(i), \mathbf{c}_1, \ldots, \mathbf{c}_n]\} \in \mathcal{U} \\ &\quad \text{iff} \quad \mathfrak{M} \models \boldsymbol{\sigma}[\alpha_1(i), \ldots, \alpha_n(i), \mathbf{c}_1, \ldots, \mathbf{c}_n] \\ &\qquad\qquad \text{and } \mathfrak{M} \models \boldsymbol{\tau}[\alpha_1(i), \ldots, \alpha_n(i), \mathbf{c}_1, \ldots, \mathbf{c}_n]\} \\ &\quad \text{iff} \quad \mathfrak{M} \models \boldsymbol{\sigma}[\alpha_1(i), \ldots, \alpha_n(i), \mathbf{c}_1, \ldots, \mathbf{c}_n] \wedge \boldsymbol{\tau}[\alpha_1(i), \ldots, \alpha_n(i), \mathbf{c}_1, \ldots, \mathbf{c}_n] \\ &\quad \text{iff} \quad \mathfrak{M} \models \boldsymbol{\theta}[\alpha_1(i), \ldots, \alpha_n(i), \mathbf{c}_1, \ldots, \mathbf{c}_n]. \quad \square \end{aligned}$$

Łoś (1955) showed the following theorem.

Theorem 11.2 (Łoś's Theorem) *Suppose*

- L *is a first order language*
- $\boldsymbol{\theta}$ *is a sentence of* L
- H *is a nonempty set*
- $\mathcal{U}$ *is an ultrafilter on* H
- *for each* i *in* H, $\langle Z_i, F_i \rangle$ *is a model of* L
- *and* $\mathfrak{M} = \langle Z, F \rangle$ *is the* $\mathcal{U}$*-ultraproduct of* $\{\mathfrak{M}_i\}_{i \in H}$.

Then

$$\mathfrak{M} \models \boldsymbol{\theta} \;\; \textit{iff} \;\; \{i \mid i \in H \textit{ and } \mathfrak{M}_i \models \boldsymbol{\theta}\} \in \mathcal{U}. \tag{11.2}$$

Proof. By Theorem 11.1, we may assume that $\boldsymbol{\theta}$ is in prenex normal form, say $\boldsymbol{\theta}$ is

$$(\exists x)(\forall y)(\exists z)(\forall u)(\forall v)(\exists w)\boldsymbol{\sigma}[x, y, z, u, v, w, \boldsymbol{a}, \boldsymbol{b}, \boldsymbol{c}].$$

Let

$$G = \{i \mid i \in H \text{ and } \mathfrak{M}_i \models \boldsymbol{\theta}\}.$$

Then for each i in G, let d_i be an element of Z_i, and φ_i and η_i be Skolem functions for $\mathfrak{M}_i$ and $\boldsymbol{\theta}$ such that

$$\begin{aligned} \mathfrak{M}_i \models \boldsymbol{\theta} &\text{ if and only if for all } y,\, u, \text{ and } v \text{ in } Z_i, \\ &\mathfrak{M}_i \models \boldsymbol{\sigma}[d_i, y, \varphi_i(y), u, v, \eta_i(y, u, v), \boldsymbol{a}, \boldsymbol{b}, \boldsymbol{c}]. \end{aligned} \tag{11.3}$$

Thus by Lemma 11.1, if $G \in \mathcal{U}$ then $\mathfrak{M} \models \boldsymbol{\theta}$. Therefore the following has been shown:

$$\text{if } \{i \in H \mid \mathfrak{M}_i \models \boldsymbol{\theta}\} \in \mathcal{U}, \text{ then } \mathfrak{M} \models \boldsymbol{\theta}. \tag{11.4}$$

If $\{i \in H \mid \mathfrak{M}_i \models \boldsymbol{\theta}\} \notin \mathcal{U}$, then it follows from Theorem 3.3 that $\{i \in H \mid \mathfrak{M}_i \models \neg\boldsymbol{\theta}\} \in \mathcal{U}$. Then, by Lemma 11.1, $\mathfrak{M} \models \neg\boldsymbol{\theta}$, that is, not $\mathfrak{M} \models \boldsymbol{\theta}$. Therefore,

$$\text{if } \{i \in H \mid \mathfrak{M}_i \models \boldsymbol{\theta}\} \notin \mathcal{U}, \text{ then not } \mathfrak{M} \models \boldsymbol{\theta}. \tag{11.5}$$

The conclusion of the theorem, Equation 11.2, follows from Equations 11.4 and 11.5. □

Definition 11.12 Let $\mathfrak{M} = \langle Z, F \rangle$ be a model of the first order language L and $\mathcal{U}$ be an ultrafilter on H. Then $\mathfrak{N}$ is said to be the $\mathcal{U}$ – *ultrapower* of $\mathfrak{M}$ if and only if $\mathfrak{N}$ is the $\mathcal{U}$-ultraproduct of $\{\mathfrak{M}_i\}_{i \in H}$, where for each i in H, $\mathfrak{M}_i = \mathfrak{M}$.

$\mathfrak{P}$ is said to be an *ultrapower* of $\mathfrak{M}$ if and only if for some ultrafilter $\mathcal{V}$, $\mathfrak{P}$ is the $\mathcal{V}$-ultrapower of $\mathfrak{M}$. □

Theorem 11.3 *Let* L *be a first order language,* $\mathfrak{M}$ *be a model of* L*, and* $\mathfrak{N}$ *be an ultrapower of* $\mathfrak{M}$*. Then for each sentence* $\boldsymbol{\theta}$ *of* L*,*

$$\mathfrak{M} \models \boldsymbol{\theta} \text{ iff } \mathfrak{N} \models \boldsymbol{\theta}\,.$$

Proof. Immediate from Theorem 11.2. □

11.5 Compactness Theorem of Logic

Definition 11.13 Let L be a first order language and Γ be a set of sentences (possibly empty) of L. Then $\mathfrak{M}$ is said to be a *model* of Γ if and only if $\mathfrak{M}$ is a model of L and for each $\boldsymbol{\theta}$ in Γ, $\mathfrak{M} \models \boldsymbol{\theta}$. Γ is said *to have a model* if and only if for some $\mathfrak{N}$, $\mathfrak{N}$ is a model of Γ. □

Theorem 11.4 (The Compactness Theorem of Logic) *Let* L *be a first order language and* Γ *be a nonempty set of sentences of* L *such that each nonempty finite subset of* Γ *has a model. Then* Γ *has a model.*

Proof. Let

$$S = \text{the set of nonempty finite subsets of } \Gamma,$$

and for each $\boldsymbol{\theta}$ in Γ, let

$$\widehat{\boldsymbol{\theta}} = \{\Delta \mid \Delta \in S \text{ and } \boldsymbol{\theta} \in \Delta\},$$

and let $\mathcal{F} = \{\widehat{\boldsymbol{\theta}} \mid \boldsymbol{\theta} \in \Gamma\}$. Then $\mathcal{F}$ has the finite intersection property, because for $\widehat{\boldsymbol{\theta}}_1, \ldots, \widehat{\boldsymbol{\theta}}_n$ in $\mathcal{F}$, $\{\boldsymbol{\theta}_1, \ldots, \boldsymbol{\theta}_n\} \in \widehat{\boldsymbol{\theta}}_i$ for $i = 1, \ldots, n$. Let $\mathcal{U}$ be an ultrafilter that contains $\mathcal{F}$. By hypothesis, for each Δ in S let $\mathfrak{M}_\Delta$ be a model of Δ. Let $\mathfrak{M}$ be the $\mathcal{U}$-ultraproduct of $\{M_\Delta\}_{\Delta \in S}$. It will be shown that $\mathfrak{M}$ is a model of Γ. Let $\boldsymbol{\theta}$ be an arbitrary sentence of Γ. Then for each Δ in $\widehat{\boldsymbol{\theta}}$, $\mathfrak{M}_\Delta \models \boldsymbol{\theta}$. Thus

$$\widehat{\boldsymbol{\theta}} \subseteq \{\Delta \in S \mid \mathfrak{M}_\Delta \models \boldsymbol{\theta}\}\,.$$

Because $\widehat{\boldsymbol{\theta}} \in \mathcal{U}$, it follows that $\{\Delta \in S \mid \mathfrak{M}_\Delta \models \boldsymbol{\theta}\} \in \mathcal{U}$, that is, $\mathfrak{M} \models \boldsymbol{\theta}$. □

Gödel (1930) proved the Compactness Theorem for countable sets of sentences, and Malcev (1936) extended it to uncountable sets of sentences. The proof of Theorem 11.4 is based on Morel, Scott, and Tarski (1958).

11.6 Löwenheim-Skolem Theorem

Definition 11.14 Let X be a set, $f_1, \ldots, f_i, \ldots$ be countably many operations on X, where for each i, f_i is an n_i-ary operation on X. Then Y is said to be the *algebraic closure of* X *under* $f_1, \ldots, f_i, \ldots$ if and only if Y is the smallest set such

(*i*) $X \subseteq Y$, and

(*ii*) for each i and each $y_1, \ldots, y_{n_i}$ in Y, $f_i(y_1, \ldots, y_{n_i})$ is in Y.

It is well-known in set theory and algebra that the algebraic closure Y of a set X under countably many operations always exists, and is countable if X is finite, and has the same cardinality as X if X is infinite. □

Definition 11.15 Throughout this book the notation "$R \restriction Y$," where R is an n-ary relation and Y is a set, stands for the *restriction of* R *to* Y, that is,

$$R \restriction Y = \{(x_1, \ldots, x_n) \mid R(x_1, \ldots, x_n) \text{ and } x_1 \in Y, \ldots, x_n \in Y\}.$$

Let $\mathfrak{M} = \langle Z, F \rangle$ be a model of the first order language L and let Y be any subset of Z such that for each individual constant symbol $\boldsymbol{c}$ of L,

$$F(\boldsymbol{c}) \in Y.$$

Let G be the function on individual constant symbols and predicates of L such that for each individual constant symbol $\boldsymbol{c}$ of L and each predicate $\boldsymbol{P}$ of L,

$$G(\boldsymbol{c}) = F(\boldsymbol{c}) \text{ and } G(\boldsymbol{P}) = F(\boldsymbol{P}) \restriction Y.$$

By definition, the *restriction* of $\mathfrak{M}$ to Y, in symbols, $\mathfrak{M} \restriction Y$ is the ordered pair $\langle Y, G \rangle$. Observe that $\mathfrak{M} \restriction Y$ is a model of L. □

Definition 11.16 Let $\mathfrak{M} = \langle Z, F \rangle$ and $\mathfrak{N} = \langle Y, G \rangle$ be models of the first order language L. Then $\mathfrak{N}$ is said to be a *submodel* of $\mathfrak{M}$ if and only if

(*i*) $Y \subseteq Z$,

(*ii*) $F(\boldsymbol{c}) = G(\boldsymbol{c})$ for each individual constant of L, and

(*iii*) $G(\boldsymbol{P}) = F(\boldsymbol{P}) \restriction Y$ for each predicate $\boldsymbol{P}$ of L. □

Definition 11.17 A first order language L is said to be *countable* if and only if L has countably many individual constant symbols and countably many predicate symbols.

Let $\mathfrak{M} = \langle Z, F \rangle$ be a model of L. Then the *cardinality* of $\mathfrak{M}$ is, by definition, the cardinality of Z. □

Theorem 11.5 Downward Löwenheim-Skolem Theorem *Suppose* L *is a countable first order language,* $\mathfrak{M} = \langle Z, F\rangle$ *is a model of* L*, and* $Y \subseteq Z$*. Then there exists a submodel* $\mathfrak{N}$ *of* $\mathfrak{M}$ *such that the following four statements are true:*

1. Y *is a subset of the domain of* $\mathfrak{N}$.
2. *If* Y *is finite, then* $\mathfrak{N}$ *is countable.*
3. *If* Y *is infinite, then* Y *and* $\mathfrak{N}$ *have the same cardinality.*
4. *For each sentence* $\boldsymbol{\theta}$ *of* L, $\mathfrak{M} \models \boldsymbol{\theta}$ *iff* $\mathfrak{N} \models \boldsymbol{\theta}$.

Proof. *Case for a Single Sentence* $\boldsymbol{\theta}$. Suppose $\boldsymbol{\theta}$ is a sentence of L such that $\mathfrak{M} \models \boldsymbol{\theta}$. By Theorem 11.1 we may assume that $\boldsymbol{\theta}$ is in prenex normal form, say,

$$\boldsymbol{\theta} = (\forall x)(\exists y)(\forall u)(\exists v)\boldsymbol{\psi}(x, y, u, v, \boldsymbol{a}),$$

where $\boldsymbol{\psi}(x, y, u, v, \boldsymbol{a})$ does not contain any quantifiers. Let $a = F(\boldsymbol{a})$ and

$$C = \{F(\mathbf{c}) \mid \mathbf{c} \text{ is an individual constant symbol of } \mathsf{L}\}.$$

Then, because L is a countable language, C is a countable set. Let ξ and η be Skolem functions such that for each d and e in Z,

$$\mathfrak{M} \models \boldsymbol{\psi}(d, \xi(d), e, \eta(d, e), a).$$

Let X be the algebraic closure of $Y \cup C$ under ξ and η (Definition 11.14). Let $\mathfrak{N} = \mathfrak{M} \restriction X$. By construction, $\mathfrak{N}$ is a submodel of $\mathfrak{M}$ such that for all d and e in X,

$$\mathfrak{N} \models \boldsymbol{\psi}(d, \xi(d), e, \eta(d, e), a),$$

from which it follows that

$$\mathfrak{N} \models (\forall x)(\exists y)(\forall u)(\exists v)\boldsymbol{\psi}(x, y, u, v, \boldsymbol{a}),$$

and thus

$$\mathfrak{N} \models \theta.$$

Thus $\mathfrak{N}$ satisfies Statements 1 and 4 of the theorem. Because C is countable, Statements 2 and 3 follow by the comment at the end of Definition 11.14 about the cardinality of the algebraic closure.

General Proof. Let

$$\mathcal{T} = \{\theta \mid \theta \text{ is a sentence of } \mathsf{L} \text{ and } \mathfrak{M} \models \theta\}.$$

By Theorem 11.1 we may assume that each $\boldsymbol{\theta}_i$ in $\mathcal{T}$ is in prenex normal form. Let

$$C = \{F(\boldsymbol{c}) \,|\, \boldsymbol{c} \text{ is an individual constant symbol of } \mathsf{L}\}\,.$$

As in the above Case for a Single Sentence, for each $\boldsymbol{\theta}_i$ in $\mathcal{T}$ we can find elements of Z and Skolem functions $\xi_1^i, \ldots, \xi_{n_i}^i$ for $\boldsymbol{\theta}_i$ that allows us to eliminate the existential quantifiers and individual constant symbols of θ_i. (For example, if θ_i is

$$(\forall x)(\exists y)(\forall u)(\exists v)\psi(x, y, u, v, \boldsymbol{a})\,,$$

as in the above Case for a Single Sentence, then $\xi_1^i = \xi$ and $\xi_2^i = \eta$, where ξ and η are as in the above Case for a Single Sentence.) Let X be the algebraic closure of $Y \cup C$ under $\xi_1^i, \ldots, \xi_{n_i}^i$, $i \in \mathbb{I}^+$. Then, because C is countable, it follows by the comment at the end of Definition 11.14 that X is countable if Y is finite, and X has the same cardinality as Y if Y is infinite. Let $\mathfrak{N} = \mathfrak{M} \upharpoonright X$. Then the above Case for a Single Sentence can be easily modified to show that $\mathfrak{N} \models \boldsymbol{\theta}_i$ for all $\boldsymbol{\theta}_i$ in $\mathcal{T}$. □

Theorem 11.6 (Löwenheim-Skolem Theorem) *Suppose* L *is a countable first order language,* Γ *is a set of sentences of* L, *and* Γ *has an infinite model. Then for each infinite cardinal* $\aleph$, Γ *has a model of cardinality* $\aleph$.

Proof. Suppose Γ has an infinite model $\mathfrak{M} = \langle Z, F\rangle$ and $\aleph$ is an arbitrary infinite cardinal. Let $\boldsymbol{S}$ be a set of *new* individual constant symbols (i.e., individual constant symbols not belonging to L) such that $\boldsymbol{S}$ has cardinality $\aleph$. Let L' be the first order language formed from L by adding the new constant symbols in $\boldsymbol{S}$ to the constant symbols of L. Thus if $\boldsymbol{C}$ is the set of constant symbols of L, then L' is the first order language that has the same predicate symbols as L and $\boldsymbol{C} \cup \boldsymbol{S}$ as its individual constant symbols. Let

$$\Sigma = \{\neg\,(\boldsymbol{a} = \boldsymbol{b}) \,|\, \boldsymbol{a} \in \boldsymbol{S} \text{ and } \boldsymbol{b} \in \boldsymbol{S}\}\,.$$

Then Σ is a set of sentences of L'. Let $\Gamma' = \Gamma \cup \Sigma$. It will be shown that each finite subset of Γ' has a model of L'.

Let Δ be an arbitrary finite subset of Γ'. Without loss of generality, suppose

$$\Delta = \{\gamma_1, \ldots, \gamma_p, \boldsymbol{\sigma}_1, \ldots, \boldsymbol{\sigma}_m\}\,,$$

where $\gamma_1, \ldots, \gamma_p$ are in Γ and $\boldsymbol{\sigma}_1, \ldots, \boldsymbol{\sigma}_m$ are in Σ. Let $\boldsymbol{s}_1, \ldots, \boldsymbol{s}_n$ be the distinct individual constant symbols that appear in some sentence in $\{\boldsymbol{\sigma}_1, \ldots, \boldsymbol{\sigma}_m\}$. Because $\mathfrak{M} = \langle Z, F\rangle$ is a model of the set of sentences Γ of the language L, $\mathfrak{M}$ is a model of the set of sentences $\gamma_1, \ldots, \gamma_p$. We will now "extend" $\mathfrak{M}$ to a model $\mathfrak{N}$ of the set of sentences Δ of L'.

Let $a_1, \ldots, a_n$ be distinct elements of Z. Let G be the function on the set of predicates of L' and individual constant symbols of L' such that

(*i*) if $\boldsymbol{P}$ is a predicate of L', then $G(\boldsymbol{P}) = F(\boldsymbol{P})$;

(*ii*) if $\boldsymbol{c}$ is a constant symbol of L, then $G(\boldsymbol{c}) = F(\boldsymbol{c})$;

(*iii*) if $\boldsymbol{s}$ is in $\boldsymbol{S} - \{\boldsymbol{s}_1, \ldots, \boldsymbol{s}_n\}$, then $G(\boldsymbol{s}) = a$, where a is some arbitrary element of Z; and

(*iv*) if $\boldsymbol{s} = \boldsymbol{s}_i$ for some $i = 1, \ldots, n$, then $G(\boldsymbol{s}) = a_i$.

Then $\mathfrak{N} = \langle N, G\rangle$ is a model of L', and by construction, $\mathfrak{N}$ is a model of Δ.

The above shows that each finite subset Δ of Γ' has a model of L'. By the Compactness Theorem of Logic, let $\mathfrak{P} = \langle W, H\rangle$ be a model of L' that is a model of Γ'. Because for all distinct $\boldsymbol{s}$ and $\boldsymbol{t}$ in $\boldsymbol{S}$,

$$\mathfrak{P} \models \neg(\boldsymbol{s} = \boldsymbol{t}),$$

$H \restriction \boldsymbol{S}$ is a one-to-one function of $\boldsymbol{S}$ onto $H(\boldsymbol{S})$ $(= \{H(\boldsymbol{s})) \mid \boldsymbol{s}) \in \boldsymbol{S}\})$, where $H(\boldsymbol{S}) \subseteq W$. Because $\boldsymbol{S}$ has cardinality $\aleph$, $H(\boldsymbol{S})$ has cardinality $\aleph$, and thus W has cardinality $\geq \aleph$. Let

$$\boldsymbol{T} = \text{the set of predicates of } \mathsf{L} \cup \text{the set of individual constant symbols of } \mathsf{L},$$

and let $K = H \restriction \boldsymbol{T}$. Then $\langle W, K\rangle$ is a model of L that is a model of Γ, where W has cardinality $\geq \aleph$. Thus it follows from the Downward Löwenheim-Skolem Theorem that there is a model of L that is a model of Γ that has cardinality $\aleph$. □

The following theorem employs a method of proof similar to that of Theorem 11.6.

Theorem 11.7 *Suppose* L *is a first order language and* Γ *is a set of sentences of* L *such that* Γ *has arbitrarily large finite models. Then* Γ *has an infinite model.*

Outline of Proof. Let $\boldsymbol{S}$ be an infinite set of new constant symbols,

$$\Sigma = \{\neg(\boldsymbol{a} = \boldsymbol{b}) \mid \boldsymbol{a} \in \boldsymbol{S} \text{ and } \boldsymbol{b} \in \boldsymbol{S}\} \text{ and } \Gamma' = \Gamma \cup \Sigma.$$

Because Γ has arbitrary large finite models, a model for each finite subset Δ of Γ' can be found by choosing a model of Γ of large enough finite cardinality so that all individual constant symbols of $\Delta \cap \Sigma$ can be interpreted as distinct elements of the domain of the model. Then by the Compactness Theorem of Logic, Γ' has a model $\mathfrak{M}$. Because each sentence in Σ is true about $\mathfrak{M}$, it follows that $\mathfrak{M}$ is infinite. □

Dedekind completeness and Achimedeanness are employed in the axiomatizations of Dedekind complete extensive structures (Definition 4.1) and extensive structures with a maximal element (Definition 4.4). These axioms are not formulated in a manner that allow for immediate translations into appropriate first order languages. It is an interesting question whether such translations exist. For Dedekind completeness the question is the following: Let L be the first order language L with its predicates consisting of $\precsim$ and $\oplus$ (and no individual constant symbols) and Γ be the set of axioms formulated in L for a Dedekind complete extensive structure except for the axiom of Dedekind completeness. (Dedekind completeness as given in Definition 1.5 was not formulated in L.) Does there exists a set of sentences Σ of L such that for each model $\mathfrak{X} = \langle X, \precsim, \oplus \rangle$ of Γ,

$$\mathfrak{X} \text{ is a model of } \Sigma \text{ iff } \langle X, \precsim \rangle \text{ is Dedekind complete?}$$

For Achimedeanness a similar question arises. For both questions the answer is "No."

Let L and Γ be as above. Suppose Σ is a set sentences of L such that each model of $\Gamma \cup \Sigma$ is Dedekind complete. A contradiction will be shown. By Theorem 4.1, each model of $\Gamma \cup \Sigma$ is isomorphic to $\langle \mathbb{R}^+, \leq, + \rangle$ and therefore has cardinality $=$ the cardinality of $\mathbb{R}^+$, which contradicts the Löwenheim-Skolem Theorem applied to $\Gamma \cup \Sigma$.

Let L be as above and Δ be the set of axioms for an extensive structure with a maximal element (Definition 4.4) except for the Archimedean axiom. Suppose Σ is a set of sentences of L such that each model of $\Delta \cup \Sigma$ is Archimedean. Then, because by Theorem 4.2, each model of $\Delta \cup \Sigma$ is isomorphic to a submodel of $\langle \mathbb{R}^+, \leq, + \rangle$, it follows that each model of $\Delta \cup \Sigma$ has cardinality $\leq$ the cardinality of $\mathbb{R}^+$, which contradicts the Löwenheim-Skolem Theorem.

Suppose Γ is a set of sentences of a first order language L and Γ has arbitrary large finite models. Then it is easy to see that there is no set of sentences Δ of L such that $\Gamma \cup \Delta$ has only finite models of Γ by applying Theorem 11.7 to $\Gamma \cup \Delta$. By letting $\Gamma = \varnothing$, it then follows that there is no set of sentences of L that has only finite models of L, i.e., "finiteness" is not first order expressible. However, the class of infinite models of L is first order expressible by the infinite set of sentences of L that says, "there are two distinct elements, there are three distinct elements, ..., there are n distinct elements, ..., etc.," for example the infinite set of sentences,

$$\{(\exists v_1)(\exists v_2)[\neg\,(v_1 = v_2)],\ (\exists v_1)(\exists v_2)(\exists v_3)[\neg\,(v_1 = v_2) \\ \wedge\ \neg\,(v_2 = v_3)\ \wedge\ \neg\,(v_1 = v_3)],\ \text{etc.}\}\,.$$

11.7 Axiomatizability and Uniqueness of Probability Representations

Quite often we know what class of models we want to axiomatize. This presents the problem of finding an appropriate first order axiomatization for the class of models. Some classes cannot be axiomatized, for example as previously shown, the class of Dedekind complete extensive structures is not first order axiomatizable. This section considers the issue of axiomatizing the qualitative variant of the Kolmogorov theory consisting of the class of qualitative probability structures with unique weak probability representations. The main result, which is due to Narens (1980), show that no finite set Σ of first order axioms can be added to the finite cancellation axioms Γ so that the models of $\Gamma \cup \Sigma$ exactly coincide with the class of qualitative probability structures with unique weak probability representations. Another result, also due to Narens (1980), shows that no infinite set Δ of first order axioms can be added Γ so that the models of $\Gamma \cup \Delta$ exactly coincide with the class of qualitative probability structures with nonunique weak probability representations.

Because the results of this section are not used elsewhere in the book, the reader may want on the first reading to skip the proofs presented in the section.

Definition 11.18 Let L be a first order language, $\mathcal{C}$ be a class of models of L, and Γ be a set of sentences of L. Then Γ is said to be a *first order axiomatization* of $\mathcal{C}$ if and only if for each model $\mathfrak{M}$ of L,

$$\mathfrak{M} \in \mathcal{C} \text{ iff } \mathfrak{M} \text{ is a model of } \Gamma.$$

$\mathcal{C}$ is said to be *first order axiomatizable* if and only if $\mathcal{C}$ has a first order axiomatization. □

Convention 11.2 Throughout the remainder of this section, let L be the first order language that has individual constant symbols $\boldsymbol{X}$ and $\boldsymbol{\varnothing}$, unary predicate $\boldsymbol{-}$, binary predicate $\precsim$, and ternary predicates U and $\mathsf{\cap}$. Interpretations of $\boldsymbol{X}$, $\boldsymbol{\varnothing}$, $\boldsymbol{-}$, $\precsim$, U and $\mathsf{\cap}$ in models of L are often denoted by X, $\varnothing$, $-$, $\precsim$, $\cup$, and $\cap$, and although $\varnothing$, $-$, $\cup$, and $\cap$ are also used respectively to denote the empty set, set complementation, set union, and set intersection, this confusion of notation should not cause much trouble, because the context will make clear which interpretation is intended. The two kinds of interpretations coincide when a boolean algebra of subsets is represented by the relational structure $\langle \mathcal{E}, \cup, \cap, -, X, \varnothing \rangle$. □

It is immediate that the definition of a "boolean algebra of subsets," Definition 2.1 is translatable into L. It then easily follows by Definitions 4.8 and 4.6 and Theorem 4.8, that the class of qualitative probability structures is first order axiomatizable. (The axiomatization uses infinitely many first order sentences, because there are infinitely many finite cancellation axioms.) □

The following example is used in the proof of Lemma 11.2 below.

Example 11.1 Let $n \in \mathbb{I}^+$, $n \geq 2$, and for $k = 0, \ldots, n-1$, let

$$A_{k+1} \text{ be the half-open interval } \left[\frac{k}{n}, \frac{k+1}{n}\right).$$

Let p be a positive integer such that $1 \leq p \leq n$, and let α be an element of A_p such that

$$\alpha \neq \frac{p-1}{n} \text{ and } \alpha \neq \frac{p-1}{n} + \frac{1}{2n}.$$

Let

$$B = \left[\frac{p-1}{n}, \alpha\right) \text{ and } C = \left[\alpha, \frac{p}{n}\right).$$

Then

$$A_1, \ldots, A_{p-1}, B, C, A_{p+1}, \ldots, A_n, \{1\}$$

is a partition of $[0,1]$. Let $\mathcal{E}$ be the boolean algebra of subsets generated by this partition, and let $\mathbb{P}$ be the restriction of Lebesgue measure on $[0,1]$ to $\mathcal{E}$. Then $\mathbb{P}$ is a probability function on $\langle X, \mathcal{E} \rangle$,

$$\mathbb{P}(A_i) = \frac{1}{n}$$

for $i = 1, \ldots, p-1, p+1, \ldots, n$, and

$$\mathbb{P}(B) = \alpha - \frac{p-1}{n} \text{ and } \mathbb{P}(C) = \frac{p}{n} - \alpha.$$

Let $\precsim$ be the binary relation on $\mathcal{E}$ such that for all E and F in $\mathcal{E}$,

$$E \precsim F \text{ iff } \mathbb{P}(E) \leq \mathbb{P}(F).$$

Let β be an arbitrary element of $\left(0, \frac{1}{n}\right)$ such that

$$\begin{aligned} \beta &\neq \alpha - \frac{p-1}{n}, \\ \beta &> \frac{1}{n} - \beta \text{ if } \mathbb{P}(B) > \mathbb{P}(C), \text{ and} \\ \beta &< \frac{1}{n} - \beta \text{ if } \mathbb{P}(B) < \mathbb{P}(C). \end{aligned}$$

Let $\mathbb{Q}$ be the function on $\mathcal{E}$ such that

(*i*) $\mathbb{Q}(\varnothing) = 0$ and $\mathbb{Q}(\{1\}) = 0$,

(*ii*) for $i = 1, \ldots, p-1, p+1, \ldots, n$,

$$\mathbb{Q}(A_i) = \frac{1}{n}, \ \mathbb{Q}(B) = \beta, \ \text{and} \ \mathbb{Q}(C) = \frac{1}{n} - \beta \, , \ \text{and}$$

(*iii*) $\mathbb{Q}(E) = \sum_{i=1}^{k} \mathbb{Q}(E_i))$, where $E = \bigcup_{i=1}^{k} E_i$ and for $i = 1, \ldots, k$,

$$E_i \in \{A_1, \ldots, A_{p-1}, B, C, A_{p+1}, \ldots, A_n, \{1\}\} \, .$$

($\mathbb{Q}$ is defined on all of $\mathcal{E}$, because each nonempty element of $\mathcal{E}$ is a finite union of the atoms $A_1, \ldots, A_{p-1}, B, C, A_{p+1}, \ldots, A_n$ of $\mathcal{E}$.) Because $A_1, \ldots, A_{p-1}, B, C, A_{p+1}, \ldots, A_n$ are the atoms of the boolean algebra of subsets $\mathcal{E}$ and $\mathbb{P}$ is a probability representation of $\langle X, \mathcal{E}, \precsim \rangle$, it easily follows that $\mathbb{Q}$ is also a probability representation of $\langle X, \mathcal{E}, \precsim \rangle$. Because

$$\mathbb{P}(B) = \alpha - \frac{p-1}{n} \neq \beta = \mathbb{Q}(B) \, ,$$

$\mathbb{P} \neq \mathbb{Q}$. □

Lemma 11.2 *There exists a weakly ordered qualitative probability structure $\mathfrak{C}$ and a sequence of weakly ordered probability structures $\mathfrak{C}_k$ such that the following four statements are true for all i and j in $\mathbb{I}^+$:*

1. *If $i < j$ then $\mathfrak{C}_i \subseteq \mathfrak{C}_j$.*
2. *$\mathfrak{C} = \bigcup_{k=1}^{\infty} \mathfrak{C}_k$.*
3. *Each $\mathfrak{C}_k$ has non-unique probability representations.*
4. *$\mathfrak{C}$ has a unique weak probability representation that takes on all rational values in $[0, 1]$.*

Proof. $\mathfrak{C}$ and $\mathfrak{C}_k$ will be constructed using Lebesgue measure on $[0, 1]$. Call two Lebesgue measurable subsets of $[0, 1]$ "equivalent" if and only if they are identical except for a set of measure 0. "Equivalent" is an equivalence relation, and it partitions the Lebesgue measurable subset of $[0, 1]$ into equivalence classes. Throughout this proof, the usual mathematical practice of ignoring sets of measure 0 is followed, although it harmlessly confuses notation a little. Thus for the purposes of this proof, the sets (0,1) and [0,1] are considered identical and the sets $[0, \frac{1}{2}]$ and $[\frac{1}{2}, 1]$ are considered disjoint.

Throughout this proof, let μ be the Lebesgue measure on $[0,1]$.
For each n and k in $\mathbb{I}^+$ such that $1 \le k \le n$, let

$$A_n^k = \text{ the closed interval } \left[\frac{k-1}{n}, \frac{k}{n}\right],$$

and let $\langle \mathcal{E}_n, \cup, \cap, [0,1], \varnothing \rangle$ be the boolean algebra generated by the Lebesgue measurable sets $A_n^1, \ldots, A_n^n$. Let $\precsim_n$ be the binary relation on $\mathcal{E}_n$ such that for all A and B in $\mathcal{E}_n$,

$$A \precsim_n B \text{ iff } \mu(A) \le \mu(B).$$

Let $\mathbb{P}_n$ be the restriction of μ to $\mathcal{E}_n$, $\mu \restriction \mathcal{E}_n$. Then

$$\mathfrak{E}_n = \langle \mathcal{E}_n, \cup, \cap, -, [0,1], \varnothing, \precsim_n \rangle$$

is a weakly ordered qualitative probability structure with probability representation $\mathbb{P}_n$. It is easy to show that

(*i*) $\mathbb{P}_n(A_n^k) = \frac{1}{n}$ for $k = 1, \ldots, n$;

(*ii*) $\mathbb{P}_n$ takes on the values and only the values $\frac{k}{n}$ for $k = 0, \ldots, n$; and

(*iii*) for $k = 1, \ldots, n$, $\mathfrak{E}_k$ is a substructure of $\mathfrak{E}_{n!}$.

Let

$$\mathcal{E} = \bigcup_{i=1}^{\infty} \mathcal{E}_{i!}, \quad \precsim = \bigcup_{i=1}^{\infty} \precsim_{i!}, \quad \text{and } \mathbb{P} = \mu \restriction \mathcal{E}.$$

For each A in $\mathcal{E}$ and n in $\mathbb{I}^+$, if $A \in \mathcal{E}_n$ then $\mathbb{P}(A) = \mathbb{P}_n(A)$. Using this fact, it is easy to verify that

$$\mathfrak{E} = \langle \mathcal{E}, \cup, \cap, [0,1], -, \varnothing, \precsim \rangle$$

is a weakly ordered qualitative probability structure with a probability representation $\mathbb{P}$. It is also easy to show that $\mathbb{P}$ takes on, and only takes on, values of the form $\frac{k}{n}$, where k is in $\mathbb{I}^+ \cup \{0\}$, $n \in \mathbb{I}^+$, and $k \le n$.

Let

- α be an irrational number such that $0 < \alpha < 1$
- $Z = [0, \alpha]$
- $\langle \mathcal{F}_n, \cup, \cap, -, [0,1], \varnothing \rangle$ be the boolean algebra of subsets generated by $\mathcal{E}_n$ and $\{Z\}$

- $\mathcal{F} = \bigcup_{i=1}^{\infty} \mathcal{F}_i$
- $\mathbb{Q}$ be the restriction of μ on $[0, 1]$ to $\mathcal{F}$
- $\precsim'$ be the binary relation $\mathcal{F}$ such that for all A and B in $\mathcal{F}$,

$$A \precsim' B \text{ iff } \mathbb{Q}(A) \leq \mathbb{Q}(B)$$

- and $\mathfrak{F} = \langle \mathcal{F}, \cup, \cap, [0, 1], \varnothing, \precsim' \rangle$.

Then it easily follows that $\mathfrak{F}$ is a weakly ordered qualitative probability structure with probability representation $\mathbb{Q}$. Let

- $\precsim'_n$ be $\precsim' \restriction \mathcal{F}_n$
- $\cup_n = \cup \restriction \mathcal{F}_n$, $\cap_n = \cap \restriction \mathcal{F}_n$, and $-_n = - \restriction \mathcal{F}_n$
- $\mathfrak{F}_n = \langle \mathcal{F}_n, , \precsim'_n, \cup_n, \cap_n, -_n, [0, 1], \varnothing, \precsim'_n \rangle$
- and $\mathbb{Q}_n = \mathbb{Q} \restriction \mathcal{F}_n$.

Then $\mathfrak{F}_n$ is a weakly ordered qualitative probability structure, and $\mathbb{Q}_n$ is a weak representation for $\mathfrak{F}_n$, and $\mathbb{Q}_n(Z)$ is the irrational number α. Furthermore,

(*i*) $\mathfrak{F}_n$ is a substructure of $\mathfrak{F}_{(n+1)!}$, in symbols, $\mathfrak{F}_n \subseteq \mathfrak{F}_{(n+1)!}$, and

(*ii*) $\mathfrak{F}$ is the union of the chain of structures $\mathfrak{F}_{n!}$, in symbols, $\mathfrak{F}_n = \bigcup_{n=1}^{\infty} \mathfrak{F}_{(n+1)!}$.

Observe that in the structure $\mathfrak{E}_n$, $A_n^k \sim_n A_n^j$ for all j and k in $\mathbb{I}^+$ such that $1 \leq j, k \leq n$, and that for all finite subsets E of $[0, 1]$ that are in $\mathcal{E}_n$, $E \sim_n \varnothing$. It easily follows from this and the method of construction of $\mathfrak{E}_n$ that all weak probability representations of $\mathfrak{E}_n$ are identical, that is, that $\mathbb{P}_n = \mathbb{P} \restriction \mathcal{E}_n$ is the unique weak probability representation of $\mathfrak{E}_n$. Because $\mathfrak{E} = \bigcup_{i=1}^{\infty} \mathfrak{E}_{i!}$ and for each n in $\mathbb{I}^+$, $\mathfrak{E}_{n!}$ is a substructure of $\mathfrak{E}_{(n+1)!}$, it follows that $\mathbb{P}$ is the unique weak probability representation of $\mathfrak{E}$.

It follows by Example 11.1 that for each n in $\mathbb{I}^+$, $\mathfrak{F}_n$ does not have a unique probability representation. However, the infinite structure $\mathfrak{F}$ does have a unique probability representation, which can be seen by observing that $\mathbb{P}$ is the unique probability representation for $\mathfrak{E}$ and $\mathbb{P}$ takes on every rational value in $[0, 1]$. Thus for each event B in $\mathcal{F}$, there exist sequences of event B_i and C_i in $\mathcal{E}$ such that $\lim_{i \to \infty}(\mathbb{P}(B_i) - \mathbb{P}(C_i)) = 0$ and $C_i \precsim' B \precsim' B_i$, and therefore it follows that for each weak probability representation $\mathbb{T}$ of $\mathfrak{F}$, $\mathbb{T}(B) = \lim_{i \to \infty} \mathbb{P}(B_i)$, and thus the uniqueness of $\mathbb{T}$ follows from the uniqueness of $\mathbb{P}$.

The Lemma follows by choosing $\mathfrak{C} = \mathfrak{F}$ and $\mathfrak{C}_k = \mathfrak{F}_{k!}$. □

Theorem 11.8 *The class of qualitative probability structures with nonunique weak probability representations is not first order axiomatizable.*

Proof. Let Δ be a set of first order axioms for the class of qualitative probability structures with nonunique weak probability representations. A contradiction will be shown.

Let $\mathfrak{C}$ and $\mathfrak{C}_k$ be as in Lemma 11.2. Let $\mathcal{U}$ be an ultrafilter on $\mathbb{I}^+$ that contains all co-finite subsets of $\mathbb{I}^+$, and let ${}^\star\mathfrak{C}$ be the $\mathcal{U}$-ultraproduct of $\{\mathfrak{C}_k\}_{k\in\mathbb{I}^+}$. Then for each $k \in \mathbb{I}^+$, $\mathfrak{C}_k$ satisfies all of the axioms in Δ, and thus by Łoś's Theorem, ${}^\star\mathfrak{C}$ has at least two distinct weak probability representations. Similarly, by Łoś's Theorem, ${}^\star\mathfrak{C}$ is weakly ordered, because, by hypothesis, $\mathfrak{C}_k$ is weakly ordered for each $k \in \mathbb{I}^+$.

Without loss of generality, let

- $\mathfrak{C} = \langle \mathcal{C}, \precsim, \cup, \cap, - \rangle$
- ${}^\star\mathfrak{C} = \langle {}^\star\mathcal{C}, {}^\star\precsim, {}^\star\cup, {}^\star\cap, {}^\star- \rangle$
- $\mathfrak{C}_k = \langle \mathcal{C}_k, \precsim_k, \cup_k, \cap_k, -_k \rangle$

and for each C in $\mathcal{C}$, let i_C be the function from $\mathbb{I}^+$ into $\mathcal{C}$ such that for all k in $\mathbb{I}^+$, $i_C(k) = C$. Let D, E, and F be arbitrary elements of $\mathcal{C}$. Because $\mathcal{C}_i \subseteq \mathcal{C}_{i+1}$ for all i in $\mathbb{I}^+$ and $\mathcal{C} = \bigcup_{i=1}^{\infty} \mathcal{C}_i$, it follows that D, E, and F are in $\mathcal{C}_j$ for all but finitely many j in $\mathbb{I}^+$. It also follows that for all but finitely many j in $\mathbb{I}^+$,

$$D \cup E = F \text{ iff } D \cup_j E = F .$$

Thus, because $\mathcal{U}$ is an ultrafilter containing the co-finite subsets of $\mathbb{I}^+$, it easily follows that the function φ from $\mathcal{C}$ into ${}^\star\mathcal{C}$ such that for each C in $\mathcal{C}$,

$$\varphi(C) = \text{the } \mathcal{U}\text{-equivalence class of which } i_C \text{ is an element}$$

is an isomorphic imbedding of $\mathcal{C}$ into ${}^\star\mathcal{C}$.

By the previous paragraph we may assume without loss of generality that $\mathfrak{C}$ is a substructure of ${}^\star\mathfrak{C}$. By hypothesis, let T be the unique weak probability representation of $\mathfrak{C}$, and let T_1 and T_2 be two distinct weak probability representations of ${}^\star\mathfrak{C}$. Because T is unique, $T = T_1 \restriction \mathcal{C}$ and $T = T_2 \restriction \mathcal{C}$. Because $T_1 \neq T_2$, let A in ${}^\star\mathcal{E}$ be such that $T_1(A) \neq T_2(A)$. Without loss of generality, suppose $T_1(A) < T_2(A)$. Because by hypothesis $\mathfrak{C}$ and $\mathfrak{C}_k$ are as in Lemma 11.2, T takes on all rational values in $[0, 1]$. Thus let B in $\mathcal{C}$ be such that

$$T_1(A) < T(B) < T_2(A) .$$

Because $T_1(B) = T(B) = T_2(B)$, it then follows that

$$T_1(A) < T_1(B) \tag{11.6}$$

and

$$T_2(B) < T_2(A)\,. \tag{11.7}$$

Because ${}^{\star}\precsim$ is a weak ordering on ${}^{\star}\mathcal{C}$ and T_1 and T_2 are weak probability representations, it follows from Equation 11.6 that $A \;{}^{\star}\!\prec B$ and from Equation 11.7 that $B \;{}^{\star}\!\prec A$, which is a contradiction. □

Theorem 11.9 *Let* $\mathbf{\Gamma}$ *be a first order axiomatization for the class of qualitative probability structures (e.g., let* $\mathbf{\Gamma}$ *be the set of finite cancellation axioms). Then there does not exist a finite set of first order axioms* $\mathbf{\Sigma}$ *such that* $\mathbf{\Gamma} \cup \mathbf{\Sigma}$ *is an axiomatization for the class of qualitative probability structures with unique weak probability representations.*

Proof. Suppose $\mathbf{\Sigma}$ is finite and $\mathbf{\Gamma} \cup \mathbf{\Sigma}$ is a first order axiomatization for the theory of qualitative probability structures with unique weak probability representations. A contradiction will be shown. Because $\mathbf{\Sigma}$ is finite, let $\mathbf{\Sigma} = \{\boldsymbol{\theta}_1, \ldots, \boldsymbol{\theta}_n\}$. Let

$$\boldsymbol{\theta} = \boldsymbol{\theta}_1 \wedge \cdots \wedge \boldsymbol{\theta}_n\,.$$

Then $\mathbf{\Gamma} \cup \{\neg\,\boldsymbol{\theta}\}$ axiomatizes the theory of qualitative probability structures with nonunique weak probability representations, which is impossible by Theorem 11.8. □

Narens (1980) makes the following comment about Theorems 11.8 and 11.9:

> The method of proof for [Theorems 11.8 and 11.9] does not rely in an essential way upon probabilistic concerns; rather it is the convergence and uniqueness/nonuniqueness properties of the sequence of structures mentioned in [Lemma 11.2] that allow the proofs to work. Thus this method of proof readily extends to other situations where lemmas analogous to [Lemma 11.2] hold.

11.8 Axiom of Choice and Probability Theory

11.8.1 Classical Results of Vitali, Hausdorff, Banach, and Ulam

The Axiom of Choice is a principle of set theory that says for each nonempty set $\mathcal{S}$ of nonempty sets, there exists a function f on $\mathcal{S}$ such that for each A in $\mathcal{S}$, $f(A) \in A$. The Axiom of Choice was formally introduced as principle of set theory by Zermelo (1904), and has generated more controversy than any axiom in the history of mathematics. It was initially opposed by many—if not most—of the outstanding mathematicians and philosophers of mathematics of the time. Today it is a common fixture in mathematics that is accepted by the vast majority of mathematicians, and is presented as a valid mathematical principle in almost all current, advanced level mathematics textbooks. In probability theory, the Axiom of Choice has produced some counter-intuitive results involving probability distributions. The first of these was a response to a problem of H. Lebesgue which is often called "Lebesgue's measure problem," which for the purposes of this section is renamed "Lebesgue's Probability Problem."

Definition 11.19 For a metric space with metric δ, two sets of points A and B are said to be *congruent* if and only if there exists a one-to-one function f from A onto B such that for all x and y in A, $\delta(x, y) = \delta[f(x), f(y)]$. For Euclidean spaces, the Euclidean metric is assumed to be used for defining congruence. □

In 1902, Lebesgue in a famous and far reaching paper on the theory of integration formulated the following problem:

Lebesgue's Probability Problem. *Does there exists a σ-additive probability function on the subsets of points of the p-dimensional unit cube, C_p, such that congruent subsets of C_p have the same probability?* □

In attempting to solve this problem, Lebesgue defined a σ-additive probability function and a σ-algebra of subsets of C_p that today are called respectively, "Lebesgue measure" on C_p and "Lebesgue measurable subsets" of C_p. He did not know whether his probability function solved his probability problem; that is, he did not know if every subset of C_p was Lebesgue measurable. In 1905, G. Vitali used the Axiom of Choice to show that Lebesgue's Probability Problem had no positive solution; that is, Vitali showed that for each p in $\mathbb{I}^+$ there exist non-Lebesgue measurable subsets of C_p.

Many mathematicians and philosophers during the first three decades of 20th century believed the Axiom of Choice would ultimately produce inconsistencies, and some saw Vitali's theorem as a harbinger of future contradictions. In 1914, F. Hausdorff used the Axiom of Choice to produce a result that seemed so counter-intuitive that some prominent mathematicians considered it sufficient grounds for the total rejection of the Axiom.

This theorem, known today as *Hausdorff's Paradox*, was the result of Hausdorff's investigation of a weakened form of Lebesgue's probability Problem. It is formulated as follows:

Hausdorff's Probability Problem. *Does there exist a finitely additive probability function on the set of subsets of the p-dimensional unit cube, C_p, such that congruent subsets of C_p have the same probability?* □

Note that the critical difference between Hausdorff's and Lebesgue's versions of the measure problem is that Hausdorff weakens σ-additivity of the measure to finite additivity.

Hausdorff was not able to solve his problem for Euclidean 1- and 2-dimensional spaces. However, for Euclidean p-dimensional space with $p \geq 3$, he showed that no finitely additive probability function existed satisfying the conditions of his problem. He accomplished this by showing that one half of a sphere was congruent to one third of the same sphere, or more precisely, that the sphere could be decomposed into four (disjoint) sets A, B, C, and D, where A, B, C, and $B \cup C$ were congruent and D was denumerable. Moore (1982) writes the following about the reception of this "paradox":

> The first to respond to Hausdorff's paradox—indeed, the first to characterize it as a paradox—was Borel, who felt quite certain that the culprit was the Axiom of Choice. In the second edition of his *Leçons sur la théorie des fonctions* [1914], Borel concluded his exposition of the paradox with a polemic against the Axiom:
>
> > If, then, we designate by a, b, c the probability that a point in S belongs to A, B, or C respectively and if we grant that the probability of a point belonging to a set E is not changed by a rotation around a diameter (this is what Lebesgue expresses by saying that two congruent sets have the same measure), one obtains the contradictory equalities: $a + b + c = 1$, $a = b$, $a = c$, $a = b + c$.
> >
> > The contradiction has its origin in the application ... of Zermelo's *Axiom of Choice*. The set A is homogeneous on the sphere; but it is at the same time a half and a third of it ... The paradox results from the fact that A *is not defined*, in the logical and precise sense

> of the word *defined.* If one scorns precision and logic, one arrives at contradictions. *(Borel 1914, 255–256)*

Deeper results analogous to Hausdorff's were obtained in 1924 by Banach and Tarski and in 1929 by von Neumann.

In 1929, Banach produced a different sort of answer to Lebesgue's probability problem: He showed, assuming the continuum hypothesis (i.e., all subsets of the reals are countable or have cardinality of the reals), that no σ-probability exists on the set of the subsets of the Euclidean p-dimensional unit cube such that the singleton sets of the points of the cube have probability 0. In 1939, S. M. Ulam produced a similar result using a condition that can be viewed as a very weakened form of the continuum hypothesis. The results of Banach and Ulam, which assume the stronger condition of σ-additivity, do not assume that congruent sets are assigned the same number by the probability function.

In summary, the results of Banach and Ulam indicate that a σ-additive probability function $\mathbb{P}$ on the unit cube C of a Euclidean space such that $\mathbb{P}(\{x\}) = 0$ for each x in C, is not defined on some subset of C. The result of Hausdorff shows the same for finitely additive probability functions $\mathbb{Q}$ for Euclidean spaces of dimension ≥ 3 such that $\mathbb{Q}(\{x\}) = 0$ for all x in C and are "uniform" in the sense that $\mathbb{Q}(A) = \mathbb{Q}(B)$ for all Euclidean congruent subsets of C. The results of Banach and Ulam does not extend to finitely additive probability functions, because Theorem 4.9 shows that for each infinite set X there exists a finitely additive probability function $\mathbb{T}$ on *all* subsets of X such that $\mathbb{T}(\{x\}) = 0$ for all x in X.

It is known within the metamathematics of set theory that Axiom of Choice is necessary for the holding of the abovementioned theorems of Vitali, Hausdorff, Banach and Tarski, von Neumann, Banach, and Ulam.

11.8.2 Implications for σ-additivity

Science, including theoretical foundational work as well as empirically based research, often finds it convenient to idealize a large finite setting as a denumerable union of an increasing sequence of finite settings. Because the union is denumerable, it is automatically ruled out as an event space for any σ-additive probability function that has singleton subsets of the elements of its domain having probability 0. This is one indication that requiring the general concept of probability to be σ-additive is overspecific for scientific applications.

Consider the case where the unit cube C of a Euclidean space is used to model the possible states of a universe and $\mathbb{P}$ is the probability function used to measure the uncertainty, where $\mathbb{P}(C) = 1$ and $\mathbb{P}(\{x\}) = 0$ for each x

in C. I see no a priori reason for excluding some subevents of C from having probability. I believe that if such exclusions are to be made, they should be based on specific assumptions about the nature of the uncertainty. Thus, by Ulam's result, such specific assumptions would then be required if $\mathbb{P}$ is assumed to be σ-additive.

If for theoretical reasons the modeling requires a 3-dimensional Euclidean based uniform probability distribution on C—as is the case in many physical applications—then by Hausdorff result, $\mathbb{P}$ will not be defined on some subevents of C. The structure of the application may require—as is usually the case in physical applications—that each subevent of interest is Lebesgue measurable. This would require the domain of $\mathbb{P}$ to be the boolean σ-algebra of the set of Lebesgue measurable subsets of C, which I consider to be a reasonable and natural generalization that is not an overspecification. Note that in this situation much is being specified—in particular, the Euclidean distance function is being given an interpretation, at least implicitly, in terms of uncertainty. The interpretation is not preserved if, for example, C is mapped in a one-to-one way onto the unit square in Euclidean 2-space.

In general, I believe that if σ-additivity is going to be employed in a situation where uncertainty has geometric properties, it is incumbent on the researcher to make clear and justify the assumptions of how uncertainty is linked to the geometry.

Plausible justifications for the σ-additivity of probability functions are rarely given, and outside physical situations, I believe that they are difficult to provide for most applications, and for many cases they cannot be given. It appears to me that a major reason for assuming σ-additivity in non-geometric settings is that mathematical theorems can then be used which assume a variant following principle: *If a sequence of random variables converges to a random variable f, then the integrals (with respect to probability measure) of the elements of the sequence converge to the integral of f.* While this principle, when provided with appropriate side conditions to guarantee the convergence of the integrals, produces powerful and beautiful results, it should not viewed as a general valid principle about probability. Instead, it should be viewed as a principle in need of additional justification to be provided by particular properties of the applied situation under consideration.

For the above reasons I believe finite additive probability functions should be taken as the general concept of probability in an axiomatic treatment of probability. Along this line, it should be noted that relative frequency approaches to probability, for example von Mises (1936), produce finitely additive probability functions.

References

Batchelder, W. H. and Bershad, N. J. (1979). The statistical analysis of a Thurstonian model for rating chess players. *Journal of Mathematical Psychology, 19,* 39–60.

Bradley, R. A. and M. E. Terry. (1952). The rank analysis of incomplete block designs. *Biometrika, 39,* 324–345.

Brenner, L., and Rottenstreich, Y. (1999). Focus, repacking, and the judgment of grouped hypotheses. *Journal of Behavioral Decision Making, 12,* 141–148.

de Finetti, B. (1937). La prévision: ses lois logiques, ses sources subjectives. *Ann. Inst. H. Poincaré, 7,* 1–68.

Ellis, B. (1966) *Basic Concepts of Measurement.* London: Cambridge University Press.

Ellsberg, D. (1961). Risk, ambiguity and the Savage axioms. *Quarterly Journal of Economics, 75,* 643–649.

Fine, T. L. (1973). *Theories of probability: an Examination of Foundations.* New York, Academic Press.

Fox, C. R. and Birke, R. (2002). Forecasting trial outcomes: Lawyers assign higher probabilities to scenarios that are described in greater detail. *Law and Human Behavior, 26,* 159–173.

Fox, C., Rogers, B., and Tversky, A. (1996). Options traders exhibit subadditive decision weights. *Journal of Risk and Uncertainty, 13,* 5–19.

Gödel, K. (1930). Die vollständigkeit der Axiome des logischen Funktionekalüls. *Monatshefte für Mathematik and Physik, 31,* 349–360.

Gödel, K. (1931). Über formal unentscheidbare Sätze per Principia Mathematica und verwandter Systeme I. *Monatshefte für Mathematik und Physik, 38,* 173–98.

Gödel, K. (1933). Eine Interpretation des intuitionistischen Aussagenkalküls. *Ergebnisse eines Mathematischen Kolloquiums, 4,* 39–40. English translation in J. Hintikka, Ed., *The Philosophy of Mathematics,* Oxford, 1969.

Hardy, G. H. (1910). *Orders of infinity, the "infinitärcalcül" of Paul Du Bois-Reymond.* Cambridge: University Press.

Heyting, A. (1930). Die Formalen Regeln der intuitionistischen Logik. *Sitzungsberichte der Preussichen Akademie der Wissenschaften,* 42–56. English translation in *From Brouwer to Hilbert : the debate on the foundations of mathematics in the 1920s,* Oxford University Press, 1998.

Hölder, O. (1901). Die Axiome der Quantität und die Lehre vom Mass. *Berichte über die Verhandlungen der Königlich Sächsischen Gesellschaft der Wissenschaften zu Leipzig, Mathematisch-Physikaliche Classe, Bd. 53,* 1–64. (Part I translated into English by J. Michell and C. Ernst, "The axioms of quantity and the theory of measurement," *Journal of Mathematical Psychology, 1996, Vol. 40,* 235–252.)

Idson, L. C., Krantz, D. H., Osherson, D., and Bonini, N. (2001). The relation between probability and evidence judgment: An extension of support theory. *Journal of Risk and Uncertainty, 22,* 227–249.

Kahneman, D., Slovic, P., and Tversky, A. (Eds.) (1982).*Judgment under Uncertainty: Heuristics and Biases.* New York: Cambridge University Press.

Kahneman, D., and Tversky, A. (1979). Prospect theory: an analysis of decision under risk. *Econometrica, 47(2),* 263–291.

Kahneman, D., and Tversky, A. (1982). Judgment of and by representativeness. In Kahneman, D., Slovic, P., and Tversky, A. (Eds.),*Judgment under Uncertainty: Heuristics and Biases.* New York: Cambridge University Press.

Keynes, J. M. (1929, 1962). *A Treatise on Probability.* London: Macmillan.

Kolmogorov A. (1932) Zur Deutung der intuitionistischen Logik. *Mathematische Zeitschrift, 35,* 58–65. English translation in P. Mancosu, Ed.,

From Brouwer to Hilbert : the debate on the foundations of mathematics in the 1920s, Oxford University Press, 1998.

Kolmogorov A. (1933) *Grundbegriffe der Wahrscheinlichkeitsrechnung.* Republished as *Foundations of the Theory of Probability* New York, Chelsea, 1946, 1950.

Koopman, B. O. (1940 a). The axioms and algebra of intuitive probability. *Ann. of Math., 41,* 269–292.

Koopman, B. O. (1940 b). The bases of probability. *Bull. Amer. Math. Soc., 46,* 763–774.

Koopman, B. O. (1941). Intuitive probability and sequences. *Ann. of Math., 42,* 169–187.

Krantz, D. H., Luce, R. D., Suppes, P., and Tversky, A. (1971). *Foundations of Measurement, Vol. I.* New York: Academic Press.

Łoś, J. (1955). Quelques remarques, théorémes, et problémes sur les classes définissables d'algèbres. In *Mathematical Interpretation of Formal Systems.* Amsterdam: North Holland, 98–113.

Luce, R. D. (1959). *Individual Choice Behavior.* New York: Wiley.

Luce, R. D. (1967) Sufficient conditions for the existence of a finitely additive probability measure. *Ann. Math. Statist., 38,* 780–786.

Luce, R. D. (2000). *Utility of Gains and Losses: Measurement-Theoretical and Experimental Approaches.* Mahwah: Lawrence Erlbaum and Associates.

Luce, R. D., Krantz, D. H., Suppes, P., and Tversky, A. (1990). *Foundations of Measurement, Vol. III.* New York: Academic Press.

Luce R. D., and Marley, A. A. J. (1969). Extensive measurement when concatenation is restricted and maximal elements may exist. In S. Morgenbesser, P. Suppes, and M. G. White (Eds.), *Philosophy, Science,, and method: essays in honor of Ernest Nagel.* New York: St. Martin's Press. 235–249.

Luce, R. D., and Narens, L. (1978). Qualitative Independence in Probability Theory. *Theory and Decision, 9,* 225–239.

Luce, R. D., and Narens, L. (1985). Classification of concatenation structures by scale type. *Journal of Mathematical Psychology, 29,* 1–72.

Macchi, L., Osherson, D., and Krantz, D. H. (1999). A note on superadditive probability judgments. *Psychological Review, 106,* 210–214.

Malcev, A. I. (1936) Untersuchungen aus dem Gebiete der Mathematischen Logik. *Mat. Sbornik 1, no. 43,* 323–335.

McKinsey, J. C. C., and Tarski, A. (1946). On closed elements in closure algebras. *Annals of Mathematics, 45,* 122–162.

Mellers, B., Hertwig, R., and Kahneman, K. (2001). Do frequency representations eliminate conjunction effects? An exercise in adversarial collaboration. *Psychological Science, 12,* 269–275.

Moore, G. H. (1982). *Zermelo's Axiom of Choice.* Heidelberg: Springer-Verlag.

Morel, A., Scott, D., and Tarski, A. (1958). Reduced products and the compactness theorem. *Notices Am. Math. Soc., 5,* 674–675.

Narens, L. (1974). Minimal conditions for additive conjoint measurement and qualitative probability. *Journal of Mathematical Psychology, 11,* 404–430.

Narens, L. (1976). Utility-uncertainty trade-off structures. *Journal of Mathematical Psychology, 13,* 296–322.

Narens, L. (1980). On qualitative axiomatizations for probability theory. *Journal of Philosophical Logic, 9,* 143–151.

Narens, L. (1985). *Abstract Measurement Theory.* Cambridge, Mass.: The MIT Press.

Narens, L. (2002). *Theories of Meaningfulness.* Mahwah: Lawrence Erlbaum and Associates.

Narens, L. (2003). A theory of belief. *Journal of Mathematical Psychology, 47,* 1–31.

Narens, L. (2005). A theory of belief for scientific refutations. *Synthese, 145,* 397-423.

Nisbet, R. E. and Wilson, T. D. (1977). Telling more than we can know: verbal reports on mental processes. *Psychological Review, 84,* 231–258.

Rasiowa, H. and Sikorski, R. (1968). *The Mathematics of Metamathematics.* Warsaw: Panstwowe Wydawn. Naukowe.

Redelmeier, D., Koehler, D., Liberman, V., and Tversky, A. (1995). Probability judgement in medicine: Discounting unspecified alternatives. *Medical Decision Making, 15,* 227–230.

Robinson, A. (1966). *Non-Standard Analysis.* Amsterdam: North-Holland.

Rottenstreich, Y., and Tversky, A. (1997). Unpacking, repacking, and anchoring: Advances in Support Theory. *Psychological Review, 104,* 203–231.

Saliĭ, V. N. (1988). *Lattices with Unique Complements.* Providence, RI: American Mathematical Society.

Savage, L. J. (1954). *The Foundations of Statistics.* New York: Wiley.

Schmieden, C., and Langwitz, D. (1958). Eine Erweiterung der Infinitesimalrechnung. *Math. Zeitschr., 69,* 1–39.

Scott, D. (1964) Measurement models and linear inequalities. *Journal of Mathematical Psychology, 1,* 233–247.

Scott, D., and Suppes, P. (1958). Foundational aspects of theories of measurement. *Journal of Symbolic Logic, 23,*, 113-128.

Sloman, S., Rottenstreich Y., Wisniewski C., Hadjichristidis, C., and Fox, C. R. (2004). Typical versus atypical unpacking and superadditive probability judgment. In press, *Journal of Experimental Psychology: Learning, Memory, and Cognition.*

Stone, M. H. (1936). The theory of representations for boolean algebras. *Trans. of the Amer. Math. Soc., 40,* 37–111.

Stone, M. H. (1937). Topological representations of distributive lattices and Brouwerian logics. *Čat. Mat. Fys. 67,* 1–25.

Suppes, P., Krantz, D. H., Luce, R. D., and Tversky, A. (1990). *Foundations of measurement, Vol. II.* New York: Academic Press.

Tversky, A. (1967). Additivity, utility, and subjective probability. *Journal of Mathematical Psychology, 4,* 175–201.

Tversky, A., and Kahneman, D. (1974). Judgment under uncertainty: Heuristics and biases. *Science, 185,* 1124–1131.

Tversky, A., and Koehler, D. (1994). Support Theory: A nonextensional representation of subjective probability. *Psychological Review, 101,* 547–567.

von Helmholtz, H. (1887). Zählen and Messen erkenntnistheoretisch betrachtet. *Philosophische Aufsätze Eduard Zeller gewidmet.* Leipzig.

von Mises, R. (1936). *Wahrscheinlichkeit, Statistik un Wahrheit,* 2nd *ed.* J. Springer. (English translation, *Probability, Statistics and Truth,* Dover, 1981.)

Wallsten, T., Budescu, D., and Zwick, R. (1992). Comparing the calibration and coherence of numerical and verbal probability judgments. *Management Science, 39,* 176–190.

Zermelo, E. (1904). Beweis, dass jede Menge wohlgeordnet werden kann (As einem an Herrn Hilbert gerichteten Briefe. *Mathematische Annalen, 59,* 514–516.

Zermelo, E. (1929). Die Berechnung der Turnierergebnisse als ein Maximumproblem der Wahrscheinlickkeitsrechnung. *Math. Zeitung, 29,* 436–460.

Index

D

E

F